Analytic Feedback System Design
An Interpolation Approach

Peter Dorato
University of New Mexico

Brooks/Cole
Thomson Learning™

Pacific Grove • Albany • Belmont • Boston • Cincinnati • Johannesburg • London • Madrid
Melbourne • Mexico City • New York • Scottsdale • Singapore • Tokyo • Toronto

Sponsoring Editor: *Bill Stenquist*
Marketing Team: *Nathan Wilbur, Christina DeVeto*
Marketing Communications: *Samantha Cabaluna*
Editorial Assistant: *Shelley Gesicki*
Production Coordinator: *Kelsey McGee*
Production Service: *WestWords, Inc.*

Permissions Editor: *Mary Kay Hancharick*
Cover Design: *Laurie Albrecht*
Art Editor: *Lisa Torri*
Interior Illustration: *Accurate Art, Inc.*
Typesetting: *WestWords, Inc.*
Cover Printing/Printing and Binding: *Webcom, Ltd.*

Printed in Canada

10   9   8   7   6   5   4   3   2   1

Library of Congress Cataloging-in-Publication Data
Dorato, Peter
    Analytic feedback system design : an interpolation approach /
Peter Dorato.
        p.   cm.
    ISBN 0-534-36917-0 (pbk.)
    1. Feedback control systems.   I. Title
TJ216.D65   1999
629.8′3—dc21   99-36811

To John G. Truxal and Dante C. Youla, two key persons who inspired me in my technical career.

Professor Truxal, an early advocate of analytic feedback design (control systems synthesis), was my dissertation advisor at the Polytechnic Institute of Brooklyn (now Polytechnic University).

Professor Youla, a colleague at the Polytechnic, is a long-time friend. Youla's technical contributions, e.g., strong stabilization, interpolation with positive-real functions, and parameterization of stabilizing compensators, form the basis of most of the theoretical results in this text.

Contents

Preface

This text is meant to be a supplement to existing introductory textbooks on *Feedback Control*. Most introductory texts on the subject present *trial-and-error* design techniques based on root-locus methods and the Nyquist stability criterion. The object of this supplement is to present an *analytical* approach to feedback design. The analytical approach selected is the *interpolation* approach, since it requires the least amount of mathematics. The term *analytic* means that the question of existence of a solution is settled, and that when a solution does exist, an algorithm is available which is guaranteed to solve the problem. This is in sharp contrast to *trial-and-error* methods where the existence of a solution is open and the trial-and-error algorithm may or may not arrive at a solution.

At the University of New Mexico we teach a unit (most of Chapters 2 and 3) of the transfer-function based analytical approach presented here, lasting about six lecture hours, in our introductory undergraduate feedback control course. This course is required for all electrical engineering majors. We also include material on the analytical approach in a second, elective, control course, with a focus on digital control design (Chapter 4) and robust design (Chapter 5). Special lectures on analytic design based on this material have been presented at a number of universities outside the United States, e.g., University of Catania, Polytechnic of Turin, University of Rome/Tre, and University of Calabria. Outside the United States, the first control course generally includes analytic design via state-space methods, which we cover in our second control course. Most programs in electrical, mechnical, and chemical engineering require an introductory control course where some material on analytic design can be included. More advanced material on analytic design, both transfer-function and state-space based, can be included in subsequent control courses. While models may differ from discipline to discipline, theory is fairly common to most disciplines.

I would like to acknowledge the useful inputs provided by the reviewers of the original manuscript: Theodore Djaferis, University of Massachusettes/Amherst; Jo Howze, Texas A&M University; Martin Kaliski, California Polytechnic State University; Michael Polis, Oakland University; and David Taylor, Georgia Institute of Technology. I am grateful for the invaluable help I received from my colleague, Chaouki Abdallah, in developing and putting together this book. I would also like to thank professors Luigi Fortuna, University of Catania; Giuseppe Menga, Polytechnic of Turin; Antonio Tornambé, University of Rome/Tre; and Luciano Carotenuto, University of Calabria, for their invitations to give lectures at their universities on the subject of *analytic design*. These lectures were very helpful in developing the subject matter presented here.

Peter Dorato

Acronyms

A/D	Analog to digital
BIBO	Bounded-input-bounded-output
D/A	Digital to analog
DT	Discrete-time
FIFO	Finite-sequence in, finite-sequence out
IMC	Internal model control
LHP	Left-half s-plane
LC	Liénard-Chipart
LQ	Linear quadratic
MPI	Multivariate polynomial inequality
p.i.p.	Parity interlacing property
PR	Positive real
RH	Routh-Hurwitz
RHP	Right-half s-plane
SBR	Strictly bounded real
SPR	Strictly positive real
YJBK	Youla-Jabr-Bongiorno-Kučera
ZOH	Zero-order hold

Analytic Feedback System Design

Chapter 1

Introduction

1.1 Analytical versus Trial-and-Error Design

One of the standard tools used for the design of feedback control systems is the *Nyquist stability criterion*. While this tool provides considerable insight into the analysis of feedback systems, it suffers from one very important limitation–that it can be used only in a *trial-and-error* way in the design of a compensator, essentially because of the complicated way magnitude and phase are related for rational transfer functions. For example, given the Nyquist diagram of an unstable plant, it is difficult to answer even the simple question, "Does a stable compensator *exist* that will stabilize the closed-loop system?" The compensator design problem is further complicated in the multivariable case where the criterion involves the computation of a determinant at each frequency point.

In contrast to *trial-and-error* techniques, *analytical design* techniques always include the following two elements:

- *Conditions for the existence of a solution*

- *An algorithm that is guaranteed to find the solution, when it exists*

While analytical techniques appear to be very appropriate for design, they do have some limitations. One important limitation is that the compensator is generally more complex than that obtained by trial-and-error methods. Another is that most analytical techniques deal

1

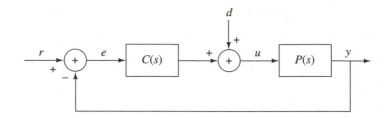

Figure 1.1: Block Diagram of Feedback System

only with limited performance measures. It is thus to the advantage of the designer to be familiar with both design techniques.

When the order of the compensator is low and fixed, a purely numerical approach is generally used to determine the compensator. Many fixed-order problems with specifications in the frequency domain can be reduced to a study of quantified *multivariate polynomial inequalities* (MPI). See reference [14]. For example, given the plant

$$P(s) = \frac{1}{s(s+1)^2}$$

and a simple proportional feedback compensator, i.e.,

$$C(s) = K$$

in a unity feedback configuration such as in Figure 1.1, if the design specification is to satisfy the tracking error specification

$$\left| \frac{1}{1 + C(j\omega)P(j\omega)} \right|^2 < 0.01, \quad \text{for all } 0 \le \omega \le 3.5 \qquad (1.1)$$

then we must meet the following bivariate polynomial inequality

$$v_1(K, \omega) = K^2 - (99 + 4K)\omega^2 - 198\omega^4 - 99\omega^6 > 0 \qquad (1.2)$$

This inequality results from clearing the fraction in the (1.1). Closed-loop stability also requires that the characteristic polynomial $s^3 + 2s^2 + s + K$ have all its zeros in the LHP. From Maxwell's condition for a cubic polynomial (see Section B.3 of Appendix B), we have the additional inequality

$$v_2(K) = 2 - K > 0 \qquad (1.3)$$

The design MPI conditions may be written compactly as

$$\forall(\omega : 0 \leq \omega \leq 3.5)[v_1(K, \omega) > 0 \ \wedge \ v_2(K) > 0] \qquad (1.4)$$

where $\forall$ denotes the logic "for-all" symbol, and $\wedge$ denotes the logic "and" symbol. The design problem is then to determine the set of design values for K that make the quantified MPI given by (1.4) true. One approach to this problem is to use *branch-and-bound techniques* (see, for example, Malan et.al. [30]) to study the validity of inequalities. Another approach is to use *Monte Carlo* type randomized algorithms (see, for example, Vidysagar [46]) to study the inequalities. These approaches are purely numerical and give no insight into the design process. However, for fixed low-order compensator design, there are very few alternatives. The number of computations required for both methods increases with the accuracy required for the final answer. We will not pursue trial and error or numeric approaches here, but the following table summarizes the advantages and disadvantages of the various design procedures.

Design Method	Advantages	Disadvantages
Analytic	Existence theorem Guaranteed convergence	Limited design objectives High-order compensator
Trail-and-error	Design insight Low-order compensator	No existence theorem No guarantee of convergence
Numeric	Multiple design objectives Low-order compensator	No design insight Computational complexity

1.2 Some Analytical Design Methods

In this book we will present one particular analytical approach to the design of compensators for feedback systems, the so-called *interpolation approach*. In the interpolation approach, various feedback design problems are converted into problems of finding special rational functions that interpolate to given values at given points in the complex s-domain. We assume here that the plant (system being controlled) can be characterized by a rational transfer function in the Laplace-transform variable s (or in the Z-transform variable z, in the discrete-time case). This approach can be introduced with a minimal amount of mathematics, e.g. Laplace-transform theory and the concept of *bounded-input-bounded-output (BIBO)* stability. We outline here some other analytical design techniques, even though most of these techniques require much more mathematics.

- *Mean-Square Design.* This approach is based on the minimization of the mean-square error

$$E\{e^2(t)\}$$

  or equivalently an integral of the form

$$\int_{-\infty}^{\infty} |E(j\omega)|^2 \, d\omega \tag{1.5}$$

  The analytical solution of the problem requires the *spectral factorization* of a polynomial, that is the factorization of a polynomial into the product of a stable polynomial and an antistable polynomial. This is the approach taken in one of the first books on analytical feedback design, i.e., the text of Newton, Gould, and Kaiser [35] published in 1957. This approach requires knowledge of stochastic processes and complex variables, but allows one to design feedback systems that can deal with random disturbances and control-effort constraints. When $E(s)$ is analytic in the right-half s-plane (RHP), the square root of the integral in (1.5) is referred to as the H^2 *norm of* $E(s)$.

- *State-Space Design.* This approach is based on a state-space representation of the plant, i.e. a representation of the form

$$\dot{x} = Ax + Bu, \quad y = Cx \tag{1.6}$$

  where x, y, and u represent the plant state, output, and input, respectively. A basic result here is that if the system is *controllable* there exists a state-feedback controller, $u(t) = -Kx(t)$, such that the poles (eigenvalues) of the closed-loop system can be located arbitrarily. State-space theory for feedback design was introduced by Kalman in the early 1960s [23]. Many textbooks are now available that focus on state-space methods (see for example [19]). In addition, most introductory control texts include at least some material on state-space design. However, this time-domain approach to analytic design requires a knowledge of matrix theory, which is not always required mathematics in engineering programs in the United States.

One state-space design theory, which is especially well suited for multivariable feedback systems, is the *linear-quadratic (LQ)* theory. In the LQ theory the problem is to find a state-feedback control law that minimizes an integral quadratic performance measure of the form

$$V = \int_0^\infty \left(x'(t)Qx(t) + u'(t)Ru(t) \right) dt$$

It can be shown that this problem can be reduced to a solution of matrix Riccati equation. See, for example, [11]. However an even more extensive knowledge of matrices is required for LQ theory. This topic is normally taught at the graduate level. In any case the transfer function approach, represented by interpolation techniques, is a useful complement to the time-domain approach represented by state-space techniques.

- H^∞ *Design.* Recently a new theory for analytical feedback system design has been evolved, based on the minimization of a performance measure of the form

$$V = \sup_\omega |E(j\omega)| \tag{1.7}$$

where, for example, $E(s)$ may be the transform of the error signal $e(t)$. When $E(s)$ is analytic in the RHP, then the value of V given in Equation (1.7) is also called the H^∞ *norm* of $E(s)$. This H^∞ theory for feedback system design was developed by Zames and Francis in 1983 [51]. An early text on the approach is that of Vidyasagar [45]. An updated treatment of the theory may be found in the recent text of Sáchez-Peña and Sznaier [40]. This theory requires rather advanced concepts in complex variables and matrices, and is generally not included in introductory courses in feedback design.

1.3 Internal Stability

A key concept in the design of feedback control systems is that of *internal stability*. Unfortunately this is a concept that is often not discussed in introductory texts. The basic idea is to insure that the closed-loop system is stable, not only between the command input and the controlled output, but also between internal points. This is important because disturbance signals can arise at all points in the closed-loop system. To define the concept, consider the feedback system shown in Figure 1.1.

Definition: The feedback system in Figure 1.1 is said to be *internally stable* if the three transfer functions:

$$E(s)/R(s) \quad = \quad \frac{1}{1 + C(s)P(s)} \tag{1.8}$$

$$Y(s)/D(s) \quad = \quad \frac{P(s)}{1 + C(s)P(s)} \tag{1.9}$$

$$U(s)/R(s) \quad = \quad \frac{C(s)}{1 + C(s)P(s)} \tag{1.10}$$

are BIBO stable.

Recall that a rational transfer function $G(s)$ is BIBO stable if and only if it is *proper* (the degree of the denominator polynomial is greater than or equal to the numerator polynomial), and its denominator polynomial is *Hurwitz* (all its roots have negative real-parts).

In the sequel we will let the "error" transfer function $E(s)/R(s)$ be denoted $S(s)$, and the "external" transfer function $Y(s)/R(s)$ be denoted $T(s)$. Note that $S(s)$ is also the *sensitivity function* for the closed-loop system, i.e.,

$$S(s) = \frac{dT/T}{dG/G} = \frac{1}{1 + C(s)P(s)} \tag{1.11}$$

Since

$$T(s) = \frac{C(s)P(s)}{1 + C(s)P(s)} = 1 - S(s) \tag{1.12}$$

the exterior transfer function $T(s)$ is often called the *complementary sensitivity function*.

Internal stability implies *external* stability, that is the stability of the transfer function $T(s) = Y(s)/R(s)$, but not conversely. For example, the PD compensator $C(s) = s + 1$ externally stabilizes the plant $P(s) = 1/s^2$, however it does not internally stabilize the closed-loop system since the transfer function

$$U(s)/R(s) = \frac{C(s)}{1 + C(s)P(s)} = \frac{s^3 + s^2}{s^2 + s + 1}$$

is not proper, hence not BIBO stable. One bad consequence of internal instability in this particular case is that a step input reference signal $r(t)$ will result in an unbounded (impulse) control input $u(t)$.

Internal stability also guarantees that there are no "bad" pole/zero cancellations between controller $C(s)$ and plant $P(s)$. "Bad cancellations" are cancellations of *unstable* poles and zeros (poles and zeros with nonnegative real parts). An example of the bad effects of unstable pole/zero cancellation is use of the compensator $C(s) = (s-1)/(s+1)$ to "stabilize" the plant $P(s) = 1/(s-1)$. This compensator does yield the stable transfer $Y(s)/R(s) = 1/(s+2)$, however it results in the unstable transfer function

$$Y(s)/D(s) = \frac{P(s)}{1 + C(s)P(s)} = \frac{(s+1)}{(s-1)(s+2)}$$

which means that a bounded disturbance signal $d(t)$ will result in an unbounded error signal $e(t)$.

1.4 The Interpolation Approach

One of the first discussions on what is now called "the interpolation approach" to control system synthesis may be found in Chapter 5 of the classic text of Truxal [44], published in 1955. The basic idea presented there (which Truxal refers to as Guillemin's method), is to use the equation

$$C(s) = \frac{1}{P(s)} \frac{T(s)}{1 - T(s)} \tag{1.13}$$

to design a compensator $C(s)$ that could be realized with an RC network. In those (pre reliable op-amp) days, RC networks were considered the most practical way to electronically realize an analog compensator. The interpolation issue arises, for example, from Equation (1.13) when one tries to avoid unstable pole zero cancellations. In particular, if the plant has an unstable zero, i.e., a zero with positive real part, then, as can be seen from Equation (1.13), the exterior transfer function $T(s)$ must "interpolate" to zero at the plant zero to avoid unstable pole/zero cancellation. This basic idea was expanded and applied to the design of digital control systems by Ragazzini and Franklin in their text [39], published in 1958. A number of texts have been published recently that develop the interpolation approach even further. See, for example, [45], [12], and [17]. One of the few introductory control texts that does include some material on the interpolation approach is the text of Bélanger [2].

In this text we will reduce various analytic feedback design problems to problems in interpolation with special functions, i.e., to problems of finding special functions, $F(s)$, which interpolate to given points, $F(\alpha_i) = \beta_i$, where generally $Re\ \alpha_i > 0$. The following table summarizes the various design problems and their corresponding interpolation problems that are treated in this text supplement.

Design Problem	Special Function, $F(s)$	Interpolation Points, α_i
Stabilization, stable $C(s)$	BIBO unit	Unstable zeros of $P(s)$
Stabilization, unstable $C(s)$	BIBO stable function	Unstable poles of $P(s)$
Gain margin design	SPR function	Unstable poles and zeros of $P(s)$
Robust stabilization	SBR function	Unstable poles of $P(s)$

Since interpolation with BIBO units and strictly-bounded-real (SBR) functions can be reduced to interpolation with strictly-positive-real (SPR) functions, the focus is on interpolation with SPR functions. With a small number of unstable poles and zeros, the SPR interpolation problem is not difficult. For an arbitrary number of points, the general algorithm of Youla and Saito [50] is presented in Appendix A.

1.5   Modeling and Numerical Examples

Since this text is meant to supplement standard introductory control texts where modeling is generally discussed in some detail, modeling is not discussed here. However since the type of plant that causes the most problems in designing feedback control systems is a plant with both poles and zeros in the RHP, we will say a few words about the model of one particular physical system, that is an *inverted pendulum* with pure delay in its control action. The transfer function obtained from a linearization about the up position of a pendulum with length L and mass M, and pure delay τ, is given by

$$P(s) = \frac{L/M}{s^2 - g/L}e^{-\tau s} \tag{1.14}$$

where g is the gravitation constant.

If the pure delay is approximated as (so-called first-order *Padé approximation*)

$$e^{-\tau s} \approx \frac{1 - \frac{\tau}{2}s}{1 + \frac{\tau}{2}s} \tag{1.15}$$

then $P(s)$ may be written

$$P(s) = L/M \frac{(2/\tau - s)}{(s + \sqrt{g/L})(s - \sqrt{g/L})(2/\tau + s)} \tag{1.16}$$

which clearly has both a pole ($s = \sqrt{g/L}$) and a zero ($s = 2/\tau$) in the RHP. Most of the "difficult" examples in this text are purely numerical, but this physical model of an inverted pendulum that generates both a pole and a zero in the RHP should be a reminder that the examples are not just academic. Finally, it should be noted that most of the examples are without physical units.

1.6 Prerequisites

It is assumed that the reader is familiar with such basic feedback techniques as root-locus, Routh–Hurwitz stability criterion, and the Nyquist stability criterion, as developed in standard introductory texts on the subject of feedback control systems. There are many introductory control texts currently in print that cover these subjects indepth. To cite a few, D'Azzo and Houpis [9], Bélanger [2], Dorf and Bishop [16], Franklin et.al. [18], Kuo [26], Marlin [31], Nise [36], Ogata [37], Phillips and Harbor [38], and Seborg et.al. [41]. Since it is assumed that this text will be used as a supplement to one of these standard texts, these basic topics are not covered in detail here. However an appendix is included (Appendix B), which summarizes some of the classical stability criteria, and the Liénard–Chipart stability criterion is introduced as an alternative to the Routh–Hurwitz criterion.

Finally, it is assumed that the reader is familiar with some kind of control systems software. Here we focus on MATLAB software, but obviously other packages may be used.

Chapter 2

Design with Stable Compensators

2.1 Introduction

In this chapter we will explore the analytical design of feedback systems using *stable* compensators. There are a number of reasons why it is important to use a stable compensator when it is possible to do so. One is that if the feedback loop is broken, the control input to the plant will remain bounded. Another is that it may then be possible to implement the compensator with a simple passive network, thus avoiding heavy and expensive power supplies. In any case, one of the first issues that arises is the question, "When can a *stable* compensator stabilize an *unstable* plant?" Stabilization with a stable compensator is now commonly referred to as *strong stabilization*. The problem of strong stabilization was solved by Youla, Bongiorno, and Lu in 1974 [48]. In [48] a very elegant necessary and sufficient condition is given for the existence of a stable stabilizing compensator, i.e. the *parity-interlacing-property (p.i.p.)*. The p.i.p. condition requires that,

between each zero of the plant on the nonnegative real axis (including infinity) in the s-domain, there be an even number of poles.

It is interesting that unstable poles or zeros off the positive real axis play no role in the question of existence of stable stabilizing compensators. Of course if the plant is stable, i.e., all poles are in the left-half

11

s-plane (LHP), and the plant is of relative degree one, then there always exists a very simple stable stabilizing compensator, that is simple proportional feedback. This is obvious from the root-locus rule, that closed-loop poles migrate from open-loop poles to open-loop zeros as gain is increased in magnitude. Thus, a stable plant can always be stabilized in a closed-loop configuration by making the gain sufficiently small. From this same rule it also follows, for a plant with relative degree equal to one, that if all the zeros of the plant are in the LHP, simple proportional control can be used to stabilize the closed-loop system by making the gain large enough. *The challenging strong stabilization problems occur when there are unstable poles and zeros on the nonnegative real axis.*

The problem of strong stabilization is discussed in some detail in Vidyasagar's text [45]. We will follow the development of the strong stabilization design approach given in [45] in our development in this chapter.

2.2   Parameterization with Units

Note that any rational transfer function, stable or otherwise, can always be written as a ratio of two *stable* transfer functions. For example, given the rational transfer function

$$P(s) = \frac{n(s)}{d(s)} \tag{2.1}$$

where $n(s)$ and $d(s)$ are arbitrary polynomials, one can always select a Hurwitz polynomial $h(s)$ such that $n(s)/h(s) = N_p(s)$ and $d(s)/h(s) = D_p(s)$ are BIBO stable functions and

$$P(s) = \frac{N_p(s)}{D_p(s)} \tag{2.2}$$

Assume that the plant transfer function $P(s)$ has been written as such a ratio, i.e. $P(s) = N_p(s)/D_p(s)$.
Now let

$$C(s) = \frac{U(s) - D_p(s)}{N_p(s)} \tag{2.3}$$

where $U(s)$ is a *unit* in the algebra of BIBO stable functions, i.e., $U(s)$ is a BIBO stable function whose inverse is also BIBO stable. Obvious necessary and sufficient conditions for a rational function to be a unit as defined here are that the numerator and denominator polynomials have the same degree, and that both polynomials be *Hurwitz*. Then the three transfer functions required in the definition of internal stability become [see Equations (1.8)–(1.10)] after substituting N_p/D_p for P, and the expression for $C(s)$ given in (2.3)

$$E(s)/R(s) \quad = \quad \frac{1}{1+C(s)P(s)} = \frac{D_p(s)}{U(s)} \tag{2.4}$$

$$Y(s)/D(s) \quad = \quad \frac{P(s)}{1+C(s)P(s)} = \frac{N_p(s)}{U(s)} \tag{2.5}$$

$$M(s)/R(s) \quad = \quad \frac{C(s)}{1+C(s)P(s)} = \frac{D_p(s)C(s)}{U(s)}. \tag{2.6}$$

In (2.6) the control input $u(t)$ in Figure 1.1 has been replaced by $m(t)$, with transform $M(s)$ to avoid confusion with the unit $U(s)$.

Let b_i denote the zeros of the plant in the RHP (including those on the $j\omega$ axis, and those at infinity), i.e. $N_p(b_i) = 0$. In order to simplify this initial discussion, we will assume that the unstable zeros of the plant are simple zeros, that is, zeros of multiplicity one. The closed-loop system will be internally stable, and the compensator will be stable, if and only if $U(s)$ is a unit that interpolates to $U(b_i) = D_p(b_i)$. This can be seen from Equations (2.4)–(2.5) by noting that since $U(s)$ is a unit, its inverse is BIBO stable; and since $U(s)$ interpolates to $D_p(s)$ at the zeros of $N_p(s)$, the compensator, given by Equation (2.3), will also be stable.

The problem of stabilization with a stable compensator has been reduced to finding a BIBO-unit $U(s)$ that interpolates to $U(b_i) = D_p(b_i)$ at the zeros, b_i, of $P(s)$ in the closed RHP.

It is obviously necessary that the interpolation values $D_p(b_i)$ at *real* interpolation points b_i all be real numbers with the *same sign*. If this were not true, then from continuity, the interpolating unit would have to go through zero on the real axis in RHP. But by definition a unit cannot be zero in the RHP! What is not so obvious is that this condition is also sufficient for the existence of an interpolating unit, and that the "same sign" condition on $D_p(b_i)$ follows from the p.i.p. condition. The details of these results may be found in Appendix A.

The unit-parameter design steps may be summarized as follows:

1. Given the plant

$$P(s) = \frac{n(s)}{d(s)}$$

   select a Hurwitz polynomial $h(s)$ of degree equal to the degree of the denominator polynomial $d(s)$, and form the stable rational functions

$$N_p(s) = \frac{n(s)}{h(s)}, \text{ and } D_p(s) = \frac{d(s)}{h(s)}$$

2. Compute a unit $U(s)$ that interpolates to

$$U(b_i) = D_p(b_i), \quad i = 1, 2, \dots, q$$

   where b_i are the q zeros of the plant $P(s)$ in the RHP including infinity.

3. The stable stabilizing compensator may then be computed from (2.3), i.e.,

$$C(s) = \frac{U(s) - D_p(s)}{N_p(s)}$$

Example 2.1. Consider two plants with transfer functions

$$P_1(s) = \frac{s-1}{(s-2)(s+2)}, \quad P_2(s) = \frac{s-2}{(s-1)(s+2)}$$

Which of these plants can be stabilized with a stable compensator?

Solution. For $P_1(s)$, between the zero at $s = 1$ and the zero at $s = \infty$ there is one pole at $s = 2$ (odd number), hence p.i.p. is not satisfied. Thus, there is no stable compensator that can stabilize this plant. For $P_2(s)$, between the zero at $s = 2$ and the zero at ∞, there are zero poles (even number), hence p.i.p. is satisfied and a stable stabilizing compensator does exist for this plant. Nyquist plots are shown in Figures 2.1 and 2.2. Both plots look very similar. It would be difficult to establish the existence, or nonexistence, of stable stabilizing compensators from these plots.

□

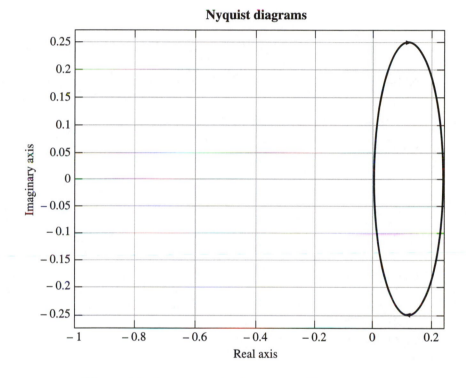

Figure 2.1: Nyquist Plot for $P_1(s)$, Example 2.1

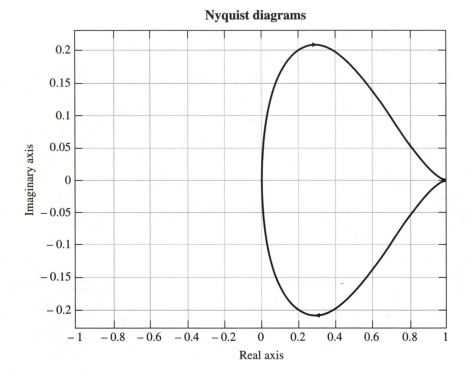

Figure 2.2: Nyquist Plot for $P_2(s)$, Example 2.1

Example 2.2. Consider the plant in Example 2.1

$$P_2(s) = \frac{s-2}{(s-1)(s+2)}$$

which can be stabilized with a stable compensator. Explore the possibility of stabilizing this plant with a zero-order compensator, i.e. simple proportional feedback $C(s) = K$. If not possible, find a dynamic stable stabilizing compensator.

Solution. From the Nyquist plot in Figure 2.2 it is clear that simple proportional feedback cannot stabilize this plant, since one can never get the correct number of encirclements of the $-1/K$ point. However, since p.i.p. is satisfied, a stable stabilizing dynamic compensator of appropriate order does exist. Root-locus analysis can also be used to show that simple proportional feedback, i.e. $C(s) = K$, cannot be used to stabilize this plant. In particular, the root-locus plot of Figure 2.3 shows that for $K > 0$ there is always a closed-loop pole in the RHP, and Figure 2.4 shows the same for $K < 0$. These plots are obtained from the MATLAB function *rlocus* (Details are given in Section B.2 of Appendix B). For stabilization with a stable compensator, at least a first-order compensator will be needed. To find a stable compensator for this second-order plant, let $h(s) = (s+2)^2$. Any second order Hurwitz polynomial would suffice. Then

$$N_p(s) = \frac{s-2}{(s+2)^2}, \quad D_p(s) = \frac{s-1}{s+2}$$

so that the unit is required to interpolate to

$$U(2) = D_p(2) = 0.25, \quad U(\infty) = D_p(\infty) = 1$$

Assume a first order unit of the form

$$U(s) = \frac{s+a}{s+b}, \quad a > 0, b > 0$$

This unit interpolates properly at $s = \infty$. To interpolate at $s = 2$, we need

$$\frac{2+a}{2+b} = 0.25$$

This interpolation condition is satisfied for all a and b that satisfy

$$b = 6 + 4a$$

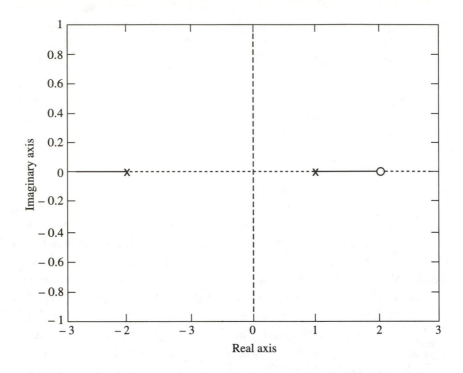

Figure 2.3: Root-locus Plot for Example 2.2, $K > 0$

If we let $a = 1$, then $b = 10$ and the required unit is given by $U(s) = (s + 1)/(s + 10)$. If this unit is substituted back into Equation (2.3), after some cancellation one obtains the compensator

$$C(s) = -6\frac{(s + 2)}{(s + 10)}$$

which is obviously stable.

□

The interpolation problem is complicated a bit if unstable zeros occur with multiplicity greater than one. In this case, in addition to the unit interpolating to given values, derivatives of the unit will have to interpolate to given values. Multiple zeros at infinity, which correspond to plants with relative degree greater than one, pose further problems. Consider first the issue of finite zeros of multiplicity greater than one. In order to simplify the discussion, let us assume in particular that the

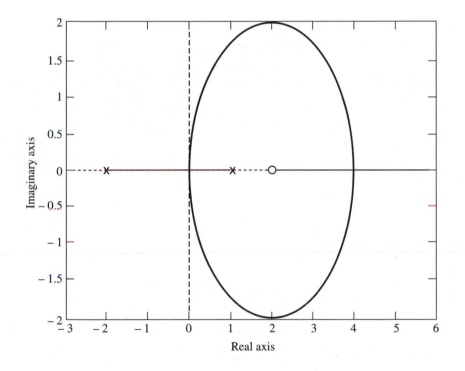

Figure 2.4: Root-locus Plot for Example 2.2, $K < 0$

plant has a finite zero at $s = b_1$ of multiplicity two. Then in order for $C(s)$ to be a stable, the interpolating unit $U(s)$ must be such that

$$U(s) - D_p(s) = (s - b_1)^2 W(s) \tag{2.7}$$

where $W(s)$ is an arbitrary function, with poles in the LHP. From (2.7) it follows, from differentiation and evaluation at $s = b_1$, that

$$U(b_1) = D_p(b_1) \text{ and } U'(b_1) = D'_p(b_1) \tag{2.8}$$

Obviously multiplicities greater than two will require interpolation conditions on higher-order derivatives of $U(s)$.

Example 2.3. Consider the plant transfer function

$$P(s) = \frac{(s - 2)^2}{s^2}$$

The problem is to find a stable stabilizing compensator for this plant.

Solution. If we select

$$N_p(s) = \frac{(s - 2)^2}{(s + 1)^2}, \quad D_p(s) = \frac{s^2}{(s + 1)^2}$$

then

$$D'_p(s) = \frac{2s}{(s + 1)^3}$$

The unit required to guarantee a stable stabilizing compensator must then satisfy the interpolation conditions

$$U(2) = D_p(2) = 4/9 \text{ and } U'(1) = D'_p(2) = 4/27$$

Consider a first-order unit

$$U(s) = \frac{as + b}{s + c}, \quad a > 0, \ b > 0, \ c > 0$$

Using the interpolation values given, after some algebra, one obtains the following for a and b:

$$a = \frac{20 + c}{27} \quad b = \frac{4c - 16}{27}$$

If we select $c = 5$, then $a = 40/27$, $b = 4/27$ and the required unit is given by

$$U(s) = \frac{40s + 4}{27(s + 5)}$$

If this unit is substituted into (2.3) we obtain the stable stabilizing compensator

$$C(s) = \frac{13s + 1}{27(s + 5)}$$

$\square$

Interpolation due to multiple zeros at infinity requires special attention. Here again, in order to simplify the discussion we will assume that the zero at infinity is of multiplicity two, that is that the plant has *relative degree* (degree of $d(s)$ minus degree of $n(s)$) two. With this assumption it follows that $N_p(s)$ can always be written

$$N_p(s) = N_2 s^{-2} + \dots \tag{2.9}$$

where ... represents terms that go to zero faster than s^{-2}, and N_2 is a nonzero constant. Then in order for $C(s)$ to be proper, the interpolating unit $U(s)$ must be such that

$$U(s) - D_p(s) = W_2 s^{-2} + \dots \tag{2.10}$$

We can satisfy (2.10) as follows. Let $w = s^{-1}$, then one can do a Taylor series expansion of $D(w)$ about $w = 0$, which corresponds to $s = \infty$, i.e.,

$$D(w) = D_0 + D_1 w + \dots \tag{2.11}$$

where ... denotes terms of order two or higher in w, and D_0 and D_1 are Taylor coefficients. Now if $U(s)$ is expanded in a similar way, i.e.,

$$U(w) = U_0 + U_1 w + \dots \tag{2.12}$$

we can satisfy (2.10) by selecting $U_0 = D_0$ and $U_1 = D_1$, but this is identical to

$$U(\infty) = D_p(\infty) \tag{2.13}$$

and

$$dU/dw = dD_p/dw, \quad w = 0 \ (s = \infty) \tag{2.14}$$

We illustrate the procedure next with a simple example.

Example 2.4. Consider the problem of strongly stabilizing the "double-integrator" plant

$$P(s) = \frac{1}{s^2}$$

Solution. Note that p.i.p. is trivially satisfied in this case since there are no finite plant zeros on the positive real axis.
Let

$$N_p(s) = \frac{1}{(s+1)^2} \text{ and } D_p(s) = \frac{s^2}{(s+1)^2}$$

then $D_p(s)$ can be written

$$D_p(s) = \frac{1}{(1+w)^2} = 1 - 2w + \ldots$$

where $w = s^{-1}$. This obviously requires $U(\infty) = 1$. A first-order unit that satisfies this condition is given by

$$U(s) = \frac{s+a}{s+b}$$

This unit may be expanded as follows

$$U(s) = \frac{1+as^{-1}}{1+bs^{-1}} = 1 + (a-b)s^{-1} + \ldots$$

Comparing the expansion of $D_p(s)$ and $U(s)$ in negative powers of s we obtain $a - b = -2$. With a arbitrarily chosen as $a = 1$, we then have $b = 3$ and the following unit

$$U(s) = \frac{s+1}{s+3}$$

If this unit is substituted back into Equation (2.3), one obtains the stable stabilizing compensator

$$C(s) = \frac{(3s+1)}{(s+3)}$$

Note that the compensator is a "lead" compensator, however no a priori assumptions were made on the structure of the compensator.

□

Complex plant zeros in the RHP pose no constraints on the existence of stable stabilizing compensators, however they do complicate the computation of interpolating units. The next example illustrates a simple case of unstable complex zeros.

Example 2.5. Consider the problem of stabilizing the plant

$$P(s) = \frac{s^2 - 2s + 2}{(s - 0.5)(s + 1)}$$

with a stable compensator.

Solution. Since this plant has no zeros on the positive real axis, including ∞, the p.i.p. condition is trivially satisfied. Let

$$N_p(s) = \frac{s^2 - 2s + 1}{(s + 1)^2}, \quad D_p(s) = \frac{s - 0.5}{s + 1}$$

then since the zeros of the plant are given by $s = 1 \pm j$, an interpolation condition is

$$U(1 + j) = D_p(1 + j) = 0.4 + 0.3j$$

If $U(s)$ is selected to be a real rational function, the interpolation condition at $s = 1 - j$ will be automatically satisfied. Assume

$$U(s) = \frac{as + b}{s + c}$$

Then evaluating $U(s)$ at the given interpolation point yields the equations

$$a = 0.7 + 0.3c, \quad b = -0.6 + 0.1c$$

Obviously both a and b can be made to have the same sign, as required for $U(s)$ to be a unit, by choosing c positive and large enough. In particular let $c = 10$, then

$$U(s) = \frac{3.7s + 0.4}{s + 10}$$

If this unit is substituted back into (2.3) we obtain the stable stabilizing compensator

$$C(s) = 2.7 \frac{(s + 1)}{(s + 10)}$$

Note that the compensator ended up being a simple phase-lead compensator, although no a priori assumptions were made, besides stability, on the structure of the compensator. The analytical interpolation theory automatically generated an appropriate compensator.

2.3   Design with Stable Compensators

We will focus our discussion here on designs based on the sensitivity function. The sensitivity function is useful for meeting error specifications and robustness conditions. Recall that the sensitivity function may be written, in terms of the unit "design parameter" $U(s)$, as

$$S(s) = \frac{D_p(s)}{U(s)} \tag{2.15}$$

The poles of $S(s)$ are obviously given by the poles of $D_p(s)$, which can be selected arbitrarily, and the zeros of $U(s)$. There is some flexibility in choosing the zeros of $U(s)$, but they must correspond to a Hurwitz polynomial and $U(s)$ must interpolate properly.

Since, from the final value theorem, one has the following expression for the steady-state error to a step input

$$e_{ss} = \frac{1}{1 + C(0)P(0)} = \frac{D_p(0)}{U(0)} \tag{2.16}$$

it can be seen that steady-state *tracking error* specifications can be reduced to an extra interpolation condition at $s = 0$, i.e.,

$$U(0) = \frac{D_p(0)}{e_{ss}} \tag{2.17}$$

if $D_p(s)$ does not have a zero at $s = 0$. Note, however, that in this case e_{ss} cannot be reduced to zero. Recall that for a type-zero plant (no integration in the plant) the compensator must have a pole at the origin to guarantee a zero steady-state error. A *stable* compensator cannot have a pole at the origin! Note also that the sign of e_{ss} must be selected so that all the $D_p(b_i)$ have the same sign.

It should be noted that problems of steady-state *disturbance rejection* can also be reduced to interpolation problems. For example, if one wishes to design a feedback system, with the structure shown in Figure 1.1, which will attenuate a unit disturbance so that the steady-state output signal $y(t)$ is reduced to a given value y_{ss}, it is clear from Equation (2.5) and the final-value theorem, that all one requires is the satisfaction of the following interpolation condition at $s = 0$

$$U(0) = \frac{N_p(0)}{y_{ss}}$$

Example 2.6. Consider a plant with transfer function

$$P(s) = \frac{(s - 2)}{(s - 1)(s + 2)}$$

This plant has a zero at $s = 2$ and a zero at $s = \infty$. Since the pole at $s = 1$ is not between these two zeros, the p.i.p. condition is satisfied and a stable stabilizing compensator is known to exist. Assume the following design specifications. The sensitivity function is to have all its poles at $s = -2$ and a steady-state error to a step input $e_{ss} = \pm 0.1$. The problem is to find a stable compensator that satisfies these specifications.

Solution. Pick $h(s) = (s + 2)^2$ to meet the pole specifications. Then

$$N_p(s) = \frac{s - 2}{(s + 2)^2}, \quad D_p(s) = \frac{s - 1}{s + 2}$$

and the interpolation conditions become

$$U(2) = D_p(2) = 0.25, \quad U(\infty) = D_p(\infty) = 1$$

To meet the steady-state error specification pick $e_{ss} = -0.1$ so that

$$U(0) = D_p(0)/e_{ss} = \frac{-0.5}{-0.1} = 5$$

This keeps the sign of all the interpolation values positive. In order to meet the pole specifications, the interpolating unit is taken to be of the form

$$U(s) = \frac{(s + 2)^2}{(s^2 + as + b)}, \quad a > 0, b > 0$$

Note that this unit already interpolates properly at $s = \infty$. By direct evaluation, the values of a and b required to meet the interpolation conditions at $s = 0$ and $s = 2$ are computed to be $a = 29.6$ and $b = 0.8$. From (2.3), the compensator is then computed to be

$$C(s) = -22.6\frac{(s+2)(s+0.195)}{s^2 + 29.6s + 0.8}$$

□

Finding units that interpolate many points is a nontrivial problem. The general problem of interpolation with units is discussed in Appendix A. In this chapter we have limited ourselves to problems that can be solved with second-order units, that is, rational functions of the form

$$U(s) = \frac{as^2 + bs + c}{s^2 + ds + e} \tag{2.18}$$

where the numerator and denominator polynomials are both Hurwitz polynomials and $a \neq 0$. The conditions for a second-order polynomial $as^2 + bs + c$ to be Hurwitz are very simple; all the coefficients must have the same sign. However, for higher-order polynomials the Hurwitz stability conditions on the coefficients become much more complex; some kind of general algorithm is required to solve the unit interpolation problem. See, for example, the algorithm given in Appendix A.

2.4 Notes and References

A general algorithm for interpolating with units is included in the original paper by Youla, Bongiorno, and Lu [48] on strong stabilization. However in reference [13] a different algorithm is given, which in general requires lower-order units. A special case, i.e., real interpolation points, of the algorithm given in [13] is included in Section A.3 of Appendix A.

It is interesting to note that the problem of *simultaneously* stabilizing two plants with one controller also reduces to interpolation with units. The simultaneous stabilization problem is important in applications where there may be sensor or actuator failures, or where a system is linearized about several operating points. More details on the two-plant stabilization problem are given in Section 5.4.

2.5 Problems

Problem 2.1. Indicate which of the following plants can be stabilized with a stable compensator.

$$P_1(s) = \frac{s^2 + s - 2}{s^3 - s^2 - 5s - 3} \tag{2.19}$$

$$P_2(s) = \frac{s^2 - 3s + 2}{s^2 - 3.9s + 2.7} \tag{2.20}$$

$$P_3(s) = \frac{s - 1}{s^2 - 6s + 9} \tag{2.21}$$

$$P_4(s) = \frac{s}{s^3 - 3s^2 + 3s - 1} \tag{2.22}$$

Problem 2.2. For the plants in problem 2.1 that can be stabilized with a stable compensator, find one. In each case, explore first the possibility of stabilization with a zero-order stable compensator, that is, simple proportional feedback. *Hint:* Root-locus or Nyquist diagram.

Problem 2.3. Consider a plant with transfer function

$$P(s) = \frac{(s - 1)(s - 2)}{(s - 0.9)(s + 1)}$$

This plant obviously satisfies the p.i.p. condition since there are no poles between the zeros at $s = 1$ and $s = 2$. Can this plant be stabilized with proportional feedback? Explore the strong stabilization of this plant using first- and second-order units.

Problem 2.4. Consider a plant with transfer function

$$P(s) = \frac{s^4 + 2s^3 - 2s - 1}{s^4 - 8s^3 + 24s^2 - 32s + 16}$$

Find a stable compensator $C(s)$ that places all the closed-loop poles at $s = -1$ and has a steady-state error of $e_{ss} = \pm 0.1$. Use the Nyquist plots of $P(s)$ and $C(s)P(s)$ to verify that the open-loop system is unstable while the closed-loop system with the designed compensator is stable.

Problem 2.5. Consider the plant

$$P(s) = \frac{(s-1)}{(s-2)^2}$$

Design a stable compensator, if one exists, that places all the poles of the sensitivity function at $s = -3$, and at the same time yields a steady-state $y_{ss} = 0.1$ for a unit step disturbance input $d(t)$.

Problem 2.6. Consider a plant with transfer function

$$P(s) = -0.164 \frac{(s+0.2)(s-0.32)}{s^2(s+0.25)(s-0.009)}$$

This plant is a model used for an automatic *ship steering* system (See problem 4.47 in [18]) with output equal to ship heading angle, and input equal to rudder angle. The p.i.p. condition is satified for this plant, so a stable stabilizing compensator exists. Find one. Note that the relative degree for this plant is two, so that a high enough order unit must be selected to meet the interpolation conditions at infinity.

Problem 2.7 Consider plants of order n that satisfy p.i.p. have only finite zeros, and r of them are are in the RHP. Show that the order of the stable compensator that stabilizes the closed-loop system is given by

$$order(C(s)) = order(U(s)) + (n-r)$$

where $U(s)$ is the required interpolating unit.

Problem 2.8 Consider the inverted-pendulum-with-delay discussed in Section 1.5, with $\tau = 2$, $L/M = g/L = 1$. The plant transfer function is then

$$P(s) = \frac{(2-s)}{(s+1)(s-1)(2+s)}$$

(a) Verify that the p.i.p. condition is satisfied for this plant.
(b) Use root-locus analysis to show that this plant cannot be stabilized with simple proportional feedback.
(c) Use unit-parameter theory to compute a stable compensator that does stabilize this plant.
Hint: Pick a high enough order unit to interpolate the zero of multiplicity two at infinity.

Chapter 3

Design with Unstable Compensators

3.1 Introduction

In this chapter we will focus on the design of compensators for systems that do not satisfy the p.i.p. condition. In this case one is forced to use an unstable compensator! However, the technique developed, which will be referred to as the *Q-parameter* design, applies even if the p.i.p. condition is satisfied. The Q-parameter approach may yield an unstable compensator even in those cases where p.i.p. is satisfied. However, the order of the compensator may be much lower than the one designed via units, since interpolating units may be of high order. Moreover, it should be noted that Q-parameter designs may lead to stable compensators.

3.2 Q-Parameter Design

If the compensator is written as

$$C(s) = \frac{Q(s)}{1 - Q(s)P(s)} \qquad (3.1)$$

then the three transfer functions involved in the definition of internal stability, see (1.8)–(1.10), become

$$E(s)/R(s) \quad = \quad 1 - P(s)Q(s) \qquad (3.2)$$

$$Y(s)/D(s) = P(s)(1 - P(s)Q(s)) \qquad (3.3)$$
$$U(s)/R(s) = Q(s) \qquad (3.4)$$

Let a_i denote the poles of $P(s)$ in the RHP (including the $j\omega$ axis). Then from Equations (3.2)–(3.4) it can be seen that internal stability requires that

1. $Q(s)$ be a BIBO stable function

2. $Q(a_i) = 0$

3. $1 - P(a_i)Q(a_i) = 0$

The interpolation conditions imposed on the stable function $Q(s)$ are a direct consequence of the internal stability constraint. For example, the term $(1 - P(s)Q(s))$ will not be stable unless $Q(s)$ is zero at an unstable pole of $P(s)$; and the term $P(s)(1 - P(s)Q(s))$ will not be stable unless $(1-P(s)Q(s))$ is zero at an unstable pole of $P(s)$. In order to simplify the discussion we will assume initially that the unstable poles are simple. If $B(s)$ is any stable function that interpolates to zero at the unstable poles of $P(s)$, i.e., $B(a_i) = 0$, then with

$$Q(s) = B(s)\tilde{Q}(s) \qquad (3.5)$$
$$\tilde{P}(s) = B(s)P(s) \qquad (3.6)$$

where $\tilde{Q}(s)$ is any BIBO stable function, it follows that the interpolation conditions can be reduced to the single interpolation condition

$$\tilde{Q}(a_i) = 1/\tilde{P}(a_i) \qquad (3.7)$$

The interpolation conditions represented in Equation (3.7) may always be satisfied with a stable function $\tilde{Q}(s)$ by fixing the denominator of $\tilde{Q}(s)$ to be an arbitrary Hurwitz polynomial,

$$d_Q(s) = s^{q-1} + \alpha_1 s^{q-2} + ... + \alpha_{q-1},$$

where q is the number of interpolation points (equal to the number of unstable poles), and then by selecting the coefficients of a numerator polynomial

$$n_Q(s) = \beta_1 s^{q-1} + \beta_2 s^{q-2} + ... + \beta_q$$

to satisfy the system of linear equations

$$\beta_1(a_i)^{q-1} + \beta_2(a_i)^{q-2} + \ldots + \beta_q = \frac{d_Q(a_i)}{\tilde{P}(a_i)} = \delta_i \ i = 1, \ldots, q \qquad (3.8)$$

It then follows that if the compensator is not constrained to be stable, any feedback stabilization problem can be reduced to interpolation with a BIBO stable function, Q(s).

The interpolating numerator polynomial $n_Q(s)$ can also be computed explicitly from *Lagrange's interpolation formula*

$$n_Q(s) = \delta_1 \frac{(s-a_2)(s-a_3)\ldots}{(a_1-a_2)(a_1-a_3)\ldots} + \delta_2 \frac{(s-a_1)(s-a_3)\ldots}{(a_2-a_1)(a_2-a_3)\ldots} + \ldots \qquad (3.9)$$

It is obvious from the form of the Lagrangian formula (3.9) that $n_Q(a_i) = \delta_i$.

The Q-parameter design steps may be summarized as follows:

1. Given a plant transfer function

$$P(s) = \frac{n_P(s)}{d_P(s)}$$

with q unstable poles, a_i, select a stable function $B(s)$ that interpolates to zero at the unstable poles of $P(s)$, i.e., $B(a_i) = 0$. To simplify the discussion we will assume that there are no poles on the $j\omega$ axis. Then if the denominator polynomial $d_P(s)$ of $P(s)$ is factored into the product $d_P(s) = d_+(s)d_-(s)$, where $d_+(s)$ represents the polynomial of unstable poles, and $d_-(s)$ represents the polynomial of stable poles, one choice for $B(s)$ is

$$B(s) = \frac{d_+(s)}{d_+(-s)} \qquad (3.10)$$

Note that if $d_+(s)$ is a polynomial with all its roots in the RHP, $d_+(-s)$ will be a polynomial with all its roots in the LHP, i.e., $B(s)$ given by (3.10) is stable.

2. Compute $\tilde{P}(s)$, and select a Hurwitz denominator polynomial $d_Q(s)$, then from (3.8) compute the polynomial interpolation points δ_i.

3. Compute the numerator polynomial by solving the linear equations (3.8), or the Lagrange interpolation formula (3.9). The required stable interpolation function $\tilde{Q}(s)$ is then given by

$$\tilde{Q}(s) = \frac{n_Q(s)}{d_Q(s)} \tag{3.11}$$

4. The stabilizing compensator may then be computed from

$$C(s) = \frac{B(s)\tilde{Q}(s)}{1 - \tilde{P}(s)\tilde{Q}(s)} \tag{3.12}$$

Feedback design based on such a stable function will be referred to as *Q-parameter design*.

We pursue next the question of the order of the compensator $C(s)$ given by (3.12). We assume that all the transfer functions involved are proper, so we can define the degree of a transfer function as the degree of the denominator polynomial. Let the degree of the plant transfer function $P(s)$ be n, and the number of unstable poles be q. With

$$B(s) = \frac{d_+(s)}{d_+(-s)}, \quad \tilde{P}(s) = \frac{n_P(s)}{d_+(-s)d_-(s)}, \quad \tilde{Q}(s) = \frac{n_Q(s)}{d_Q(s)} \tag{3.13}$$

the expression for $C(s)$ given by (3.12) becomes, after cancellations and clearing of fractions,

$$C(s) = \frac{d_+(s)n_Q(s)d_-(s)}{d_+(-s)d_-(s)d_Q(s) - n_P(s)n_Q(s)} \tag{3.14}$$

The degree of denominator $C(s)$ in (3.14) is determined by the degree of the denominator term $d_+(-s)d_-(s)d_Q(s)$ and any cancellation that occurs between numerator and denominator in (3.14). By adding all the degrees of the polynomials in this term, one obtains for the degree of this term

$$(q) + (n - q) + (q - 1) = n - 1 + q$$

since degree of $d_+(-s)$ is equal to q, the degree of $d_-(s)$ is equal to $n - q$, and the degree of $d_Q(s)$ is equal to $q - 1$. But, by the interpolation condition $1 - \tilde{P}(a_i)\tilde{Q}(a_i) = 0$, the numerator term $d_+(s)$ in (3.14) must divide the denominator of $C(s)$, so that the degree of $C(s)$ is reduced by the degree of $d_+(s)$, which is q. The final result is that the degree of the compensator $C(s)$ is equal to $n - 1$. There may be additional serendipidous cancellations in (3.14) so that the final order of $C(s)$ may be less that $n - 1$. The final result is that:

Every plant $P(s)$ of order n can be feedback stabilized with a compensator $C(s)$ of order not greater than $n - 1$.

This result is so central to linear feedback system design that it could well be labeled **the Fundamental Theorem of Feedback Control**. Note that this result is not true if we are constrained to use a *stable* compensator.

Example 3.1. Consider a plant with transfer function

$$P(s) = \frac{s - 1}{(s - 2)(s + 1)}$$

The problem is to find a compensator that stabilizes this plant.

Solution. This plant does not satisfy the p.i.p. condition. However, it is possible to design an unstable compensator via Q-parameterization theory. Let

$$B(s) = \frac{s - 2}{s + 2}$$

Then

$$\tilde{P}(s) = \frac{s - 1}{(s + 2)(s + 1)}$$

and the interpolation conditions becomes

$$\tilde{Q}(2) = 1/\tilde{P}(2) = 12$$

For a single interpolation point one may use

$$\tilde{Q}(s) = 12$$

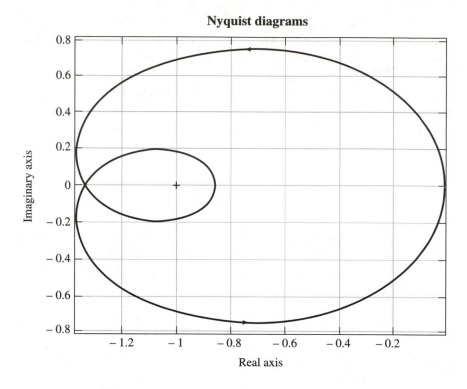

Figure 3.1: Nyquist Plot for Example 3.1

If this function is substituted back into Equations (3.5) and (3.1), one obtains, after some cancellation, the compensator

$$C(s) = 12\frac{(s+1)}{(s-7)}$$

Note that, as expected the compensator is *unstable*, and of order one, since the plant is of order two. Note also, that in computing $C(s)$ from (3.12), there is always cancellation of common unstable factors in $B(s)$ and $1 - \tilde{Q}(s)\tilde{P}(s)$. The Nyquist plot of $C(s)P(s)$ shown in Figure 3.1 verifies that the feedback system is indeed closed-loop stable. This plot was obtained from the MATLAB function *nyquist*. Details on the use of this function and the stability interpretation of the Nyquist plot may be found in Example B.1 in Appendix B.

□

The "Q-parameter" function may be further selected to meet transient and steady-state specifications. For example, from

$$E(s)/R(s) = \frac{1}{1 + C(s)P(s)} = (1 - Q(s)P(s)) \qquad (3.15)$$

it is clear that the stable poles of the error transfer function can be placed where desired by choosing the denominator of $Q(s)$, and by including in the numerator of $Q(s)$ factors that cancel undesired poles, both stable and unstable, of $P(s)$. In addition, a zero steady-state error to a step input translates to the following interpolation condition at $s = 0$:

$$\tilde{Q}(0) = \frac{1}{\tilde{P}(0)} \qquad (3.16)$$

This follows directly from the final-value theorem and the resulting equation $e_{ss} = 1 - \tilde{Q}(0)\tilde{P}(0)$.

Example 3.2. Consider the plant

$$P(s) = \frac{s - 2}{(s - 1)(s - 3)}$$

The problem is to design a compensator that stabilizes this plant, in particular places all the poles of the error transfer function in the region $Re\ s \leq -2$, and has a zero steady-state tracking error.

Solution. This plant does not satisfy the p.i.p. condition, so the compensator must be unstable. Let

$$B(s) = \frac{(s - 1)(s - 3)}{(s + 2)^2}$$

and $d(s) = (s + 2)^2$ so that

$$\tilde{Q}(s) = \frac{\beta_1 s^2 + \beta_2 s + \beta_3}{(s + 2)^2}$$

A second order $\tilde{Q}(s)$ was selected because there are *three* interpolation conditions to be met, two result from stability requirements, and the third results from the zero steady-state requirement. Then

$$\tilde{P}(s) = B(s)P(s) = \frac{s - 2}{(s + 2)^2}$$

and the error transfer function will have all its poles at $s = -2$. The interpolation conditions become

$$
\begin{aligned}
\tilde{Q}(0) &= 1/\tilde{P}(0) = -2 \\
\tilde{Q}(1) &= 1/\tilde{P}(1) = -9 \\
\tilde{Q}(3) &= 1/\tilde{P}(3) = 25
\end{aligned}
$$

The values of β_i are then determined as solutions of the interpolation equations

$$
\begin{aligned}
\beta_3 &= d(0)\tilde{Q}(0) = -8 & (3.17) \\
\beta_1 + \beta_2 + \beta_3 &= d(1)\tilde{Q}(1) = -81 & (3.18) \\
9\beta_1 + 3\beta_2 + \beta_3 &= d(3)\tilde{Q}(3) = 625 & (3.19)
\end{aligned}
$$

The solutions are $\beta_1 = 142, \beta_2 = -215$, and $\beta_3 = -8$. The appropriate Q-parameter function is then

$$
Q(s) = \frac{(s-1)(s-3)(142s^2 - 215s - 8)}{(s+2)^4}
$$

The compensator may then be computed by substituting this $Q(s)$ back into Equation (3.1). The result is

$$
C(s) = \frac{142s^2 - 215s - 8}{s(s - 130)}
$$

Note that the compensator has a pole at the origin, as expected for a zero steady-state error. The compensator is second order, which is the same order of the plant. This does not violate the fundamental theorem of feedback control because we are requiring more than just stability of the closed-loop system.

□

If a plant has an unstable pole of multiplicity greater than one, say two at $s = a_1$, then for the term $P(s)(1 - P(s)Q(s))$ to be stable we need to select $\tilde{Q}(s)$ so that

$$
1 - P(s)Q(s) = 1 - \tilde{P}(s)\tilde{Q}(s) = (s - a_1)^2 W(s) \qquad (3.20)
$$

where $W(s)$ is an arbitrary stable function. But by differentiation and evaluation at $s = a_1$, this implies

$$\tilde{P}(a_1)\tilde{Q}(a_1) = 1 \tag{3.21}$$

and that for $s = a_1$

$$\frac{d}{dt}(1 - \tilde{P}(s)\tilde{Q}(s)) = \frac{d}{dt}(\tilde{P}(s)\tilde{Q}(s)) = 0 \tag{3.22}$$

From (3.21) and (3.22), it then follows that the derivative of $\tilde{Q}(s)$ must satisfy the interpolation condition

$$\tilde{Q}'(a_1) = -\frac{\tilde{P}'(a_1)}{\tilde{P}^2(a_1)} \tag{3.23}$$

The next example illustrates the previous theory for nonsimple poles.

Example 3.3. A physical example of a control problem is given in reference [47], which is considered to be a sufficiently challenging problem to be labeled *benchmark problem*. The problem is to control two masses connected by a spring (two-mass/spring problem). The equations of motion are given by

$$M_1\ddot{x}_1 = -k(x_1 - x_2) + F, \quad M_2\ddot{x}_2 = -k(x_2 - x_1)$$

where x_1 and x_2 are the displacements of the masses M_1 and M_2, k is the spring constant, and F is a force applied to mass M_1. Consider the problem of controlling the position x_2 with the control input equal to the force F. The transfer function computed from these equations, relating input F to output x_2, is given by

$$P(s) = \frac{k}{s^2(M_1 M_2 s^2 + k(M_1 + M_2))}$$

Assume the nominal values $M_1 = M_2 = 1, k = 2$. Then the transfer function becomes

$$P(s) = \frac{2}{s^2(s^2 + 4)}$$

The problem is to find a feedback compensator that places all the closed-loop poles of the sensitivity function in the region $Re\ s \leq -1$.

Solution. This plant has a pole of multiplicity two at $s = 0$, and simple poles at $s = \pm 2j$. The p.i.p. condition is satisfied for this plant, but we will use Q-parameterization to obtain a solution that is not necessarily stable. Since $B(s)$ must be zero at the unstable poles of $P(s)$, we select

$$B(s) = \frac{s^2(s^2 + 4)}{(s + 1)^4}$$

The denominator is selected to satisfy the closed-loop pole specifications. From $B(s)$ we compute

$$\tilde{P}(s) = \frac{2}{(s + 1)^4}$$

and from $\tilde{P}(s)$ we compute the interpolation condition at the simple pole $s = 2j$ for the stable interpolating function $\tilde{Q}(s)$, i.e.,

$$\tilde{Q}(2j) = \frac{1}{\tilde{P}(2j)} = \frac{-7 - 24j}{2}$$

The pole at $s = 0$ has multiplicity two, so we need to interpolate from (3.23),

$$\tilde{Q}(0) = -\frac{\tilde{P}'(0)}{\tilde{P}^2(0)} = 2$$

in addition to

$$\tilde{Q}(0) = \frac{1}{\tilde{P}(0)} = \frac{1}{2}$$

Here the derivative of $\tilde{P}(s)$ is computed from $\tilde{P}'(s) = -8(s+1)^{-5}$. The following function should be adequate to satisfy the given interpolation conditions

$$\tilde{Q}(s) = \frac{as^3 + bs^2 + cs + d}{(s + 1)^3}$$

Note that the denominator in $\tilde{Q}(s)$ has been selected to meet the closed-loop pole specifications. Also interpolation at $s = -2j$ is not necessary since $\tilde{Q}(s)$ is a real function, and values at complex conjugate points are automatically interpolated. With this $\tilde{Q}(s)$ we then have the conditions

$$\tilde{Q}(0) = d = 1/2, \quad \tilde{Q}'(0)c - 3/2 = 2$$

and

$$\tilde{Q}(2j) = \frac{a(2j)^3 + b(2j)^2 + c(2j) + d}{(2j+1)^3)} = \frac{-7 - 24j}{2}$$

Equating real and imaginary parts, we obtain $a = -33/2$, $b = -7/2$, $c = 7/2$, $d = 1/2$. The stable interpolating function becomes

$$\tilde{Q}(s) = \frac{-33s^3 - 7s^2 + 7s + 1}{2(s+1)^3)}$$

From $B(s)$ and $\tilde{Q}(s)$ computed here we obtain [From Equation (3.12)] the compensator

$$C(s) = \frac{(-33s^3 - 7s^2 + 7s + 1)}{(s^3 + 7s^2 + 17s + 7)}$$

The compensator turns out to be stable. It would be interesting to try Nyquist of root-locus based trial-and-error procedures to meet the given stabilization specifications. Since $P(s)$ is an even function of s, a Nyquist plot of the plant is difficult to see since it coincides with the real axis on a MATLAB printout.

3.3   Youla Parameterization

Once one stabilizing compensator is found it is possible to parameter-ize all stabilizing compensators in terms of an arbitrary BIBO stable function. For example, if $\tilde{Q}_P(s)$ is a particular function that stabilizes a given plant $P(s)$, then any $\tilde{Q}(s)$ of the form

$$\tilde{Q}(s) = \tilde{Q}_P(s) + B(s)\tilde{Q}_Y(s) \tag{3.24}$$

will also stabilize the given plant, where $\tilde{Q}_Y(s)$ is an arbitrary BIBO stable function (with no interpolation constraints). This follows from the fact that

$$1 - P(s)Q(s) = (1 - \tilde{P}\tilde{Q}_P(s)) + B(s)\tilde{Q}_Y(s) \tag{3.25}$$

Note that the term $(1 - \tilde{P}(s)\tilde{Q}_P(s))$ is zero by choice of interpolation points, and the term $B(s)\tilde{Q}_Y(s)$ is zero at interpolation points for *any*

$\tilde{Q}_Y(s)$ (since $B(s)$ is zero at interpolation points), so that the conditions for internal stability are satisfied for all BIBO stable $\tilde{Q}_Y(s)$. We refer to this parameterization of all stabilizing compensators as *Youla parameterization* [49].

From Equation (3.25) it can be shown that all the transfer functions involved in internal stability, i.e., (3.2)–(3.3), may be expressed as *affine functions* (that is linear functions plus a constant) in the Youla parameter function $\tilde{Q}_Y(s)$. For example,

$$E(s)/R(s) = S(s) = (1 - P(s)Q(s)) = T_1(s) + T_2(s)\tilde{Q}_Y(s) \quad (3.26)$$

where, from (3.25)

$$T_1(s) = (1 - \tilde{P}(s)\tilde{Q}_P(s)), \quad T_2(s) = B(s) \quad (3.27)$$

Since the norm of any affine function in $\tilde{Q}_Y(s)$ is a convex function in $\tilde{Q}_Y(s)$, Youla paramaterization can be used to minimize various error norms, such as the H^2 and H^∞ norms discussed in Chapter 1. Convex functions are important in minimization problems because they have unique minimal values. The use of Youla parameterization and convex optimization is explored in detail in Boyd and Barrett [5], but is beyond the scope of this introductory text.

3.4   Internal Model Control

Note that if the plant $P(s)$ is stable, a Q-parameterized controller of the form given in (3.1) requires no interpolation conditions. This can be seen from the three transfer functions given by (3.2)–(3.4), since when $P(s)$ is stable no cancellations are required for $Q(s)$ to make these transfer functions stable. Any stable $Q(s)$ will do! In this case the controller $C(s)$ can be implemented as shown in Figure 3.2. This structure is commonly referred to as *Internal Model Control (IMC)*. See, for example, Chapter 3 of reference [34]. In this configuration $M(s)$ is a model of the plant, and $w(t)$ is a feedback signal that is nonzero only when there is a mismatch between the plant and the plant model, hence the term *internal model control*. Given a nominal transfer function $P(s)$ for the plant, then $M(s) = P(s)$. If this condition is satisfied it is easy to verify that the structure of Figure 1.1 is identical to that of

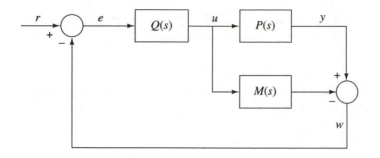

Figure 3.2: Block Diagram of IMC Feedback Structure

Figure 3.2, when $C(s)$ is given by Equation (3.1). In particular, note that with $M(s) = P(s)$, we have from the block diagram in Figure 3.2

$$u = Q(s)[r - (y - P(s)u)]$$

which becomes

$$u = \frac{Q(s)}{1 - P(s)Q(s)}(r - y)$$

which is exactly the expression for u obtained from the block diagram of Figure 1.1 when $C(s)$ is given by the Q-parameterization of Equation (3.1). For the IMC configuration of Figure 3.2, the "compensator" becomes directly the transfer function $Q(s)$, with closed-loop stability guaranteed for any $Q(s)$ that is stable.

One important advantage of the IMC feedback structure is that if the plant has a pure delay, closed-loop stability is preserved for any stable $Q(s)$, even if $Q(s)$ is restricted to being rational. Of course the plant delay must be known and incorporated into the transfer function $M(s)$. Controllers for plants with pure delay, with the structure shown in Figure 3.2, were originally referred to as *Smith-predictor* controllers. See, for example, reference [34]. The next example illustrates a Smith-predictor controller design.

Example 3.4. Consider a second-order pure time-delay plant with transfer function

$$P(s) = \frac{e^{-s}}{(s+1)^2}$$

The problem is to design a Smith-predictor controller for this plant such that the external transfer function has all its poles in the region $Res \leq -3$.

Solution. Since the external transfer function $T(s)$ may be written [recall that $T(s) = 1 - S(s)$]

$$T(s) = P(s)Q(s)$$

a simple choice for $Q(s)$ is

$$Q(s) = \frac{(s+1)^2}{(s+3)^2}$$

The Smith-predictor controller then has the structure shown in Figure 3.2 with this computed value of $Q(s)$ and $M(s) = P(s)$. Note that the external tranfer function is then given by

$$T(s) = \frac{e^{-s}}{(s+3)^2}$$

which has its poles located in the desired region, but also has a one-second delay in response, inherited from the plant delay.

3.5   Notes and References

The concept of parameterization of all stabilizing compensators was apparently independently developed by Youla, Jabr, and Bongiorno, in the journal article [49] and by Kučera in the conference paper [27]. This parameterization is commomly referred to as the *Youla parameterization* [40] , as we have done here, but it is also referred to in the literature as *Youla-Kučera parameterization* [1] , *Q-parameterization* [5], and *YJBK parameterization* [24]. In reference [49] Q-parameterization is applied to the optimal H^2 problem of continuous-time multivariable systems. For multivariable systems, the interpolation approach is awkward, and in [49] Q-parameterization is achieved by use of the so-called "Bezout identity." The theory required for multivariable systems is beyond the scope of this text. The interested reader is referred to the original paper by Youla et. al. [49] for details. Detailed discussions of the application of compensator parameterization to discrete-time systems appears in the text of Kučera [28], first published in 1979. A very comprehensive treatment of multivariable feedback system design via compensator parameterization may be found in the text of Vidyasagar[45].

In reference [7], a method very closely related to the interpolation approach is presented. It is referred to as the *linear algebraic method*. In this approach design specifications are met by matching coefficients of polynomial equations, and solving the resulting linear algebraic equations. As may be seen from Equation (3.8), Q-parameter interpolation conditons also reduce to the solution of a system of linear algebraic equations. The two methods, Q-parameter and linear algebraic, are compared in reference [8].

Detailed discussions of internal model control and Smith-predictor control may be found in the text of Morari and Zafiriou [34].

3.6 Problems

Problem 3.1. Design a stabilizing compensator for those plants in problem 2.1 that can only be stabilized by an unstable compensator.

Problem 3.2. Consider the "double-integrator" plant

$$P(s) = 1/s^2$$

This plant satisfies the p.i.p. condition, so that either U or Q parameter theory may be used to find a stabilizing compensator. Find a stabilizing compensator using Q-parameter theory, and compare the result with the compensator found in Example 2.4.

Problem 3.3. The linearized motion of an inverted pendulum with a delay in control action (applied torque) is given by the transfer function (See Section 1.5 in Chapter 1)

$$P(s) = \frac{e^{-\tau s}}{s^2 - 1}$$

Assume the pure time-delay may be approximated as follows

$$e^{-\tau s} = \frac{2/\tau - s}{2/\tau + s}$$

and that $\tau = 2.5$. Design a compensator that stabilizes the closed-loop system, with all poles in the region $Re\ s \leq -1$, and has a steady-state error (to a step reference input) of $e_{ss} = \pm 0.1$. Is it possible to design a stable compensator to meet these specifications?

Problem 3.4. Consider the inverted pendulum system in problem 3.3 but with $\tau = 1$. Is it possible to stabilize the system with a stable compensator now? Find a stabilizing compensator, not necessarily stable, for this value of delay that meets the specification in problem 3.3. What interpolation conditions on the unit $U(s)$ are required to produce a stable stabilizing compensator?

Problem 3.5. Consider the problem of stabilizing the plant

$$P(s) = \frac{1}{s^2(s^2 - 1)}$$

Use Q-parameter theory to find a stabilizing compensator. This problem appears in reference [18], problem 6.46, where one is asked to design a compensator that meets a phase margin requirement of 30 degrees.
(a) What is the phase margin of your Q-parameter design?
(b) How can the phase margin of the Q-parameter design be improved? Find a compensator that is stabilizing and meets the 30 degree phase margin requirement.
(c) Does there exist a *stable* compensator that stabilizes this plant?
(d) Show a Nyquist plot for this plant. Does the Nyquist plot provide any insight in designing a stabilizing compensator for this plant?

Problem 3.6. Consider a stable plant with transfer function

$$P(s) = \frac{e^{-2s}}{(s+1)^3}$$

Find a Smith-predictor controller that has a zero steady-state tracking error for a unit step reference input, and places all the poles of the external transfer function $T(s)$ in the region $Re\ s \leq -4$. Compute and sketch the output $y(t)$ and the control input $u(t)$ when the given Smith-predictor controller is implemented.

Problem 3.7. Consider the plant

$$P(s) = \frac{s^2 - 0.25}{s^2(s^2 - 1)}$$

Can this plant be stabilized by a *stable* compensator? Find a compensator $C(s)$ that places all the poles of the sensitivity function at $s = -2$. Show Nyquist plots of $P(s)$ and $C(s)P(s)$ to demonstrate that your compensator does stabilize the closed-loop system.

Problem 3.8. Consider the benchmark problem of Example 3.3, with the nominal values $k = M_1 = M_2 = 1$.
(a) Find a compensator that places all the closed loop poles at $s = -1$.
(b) Is this compensator stabilizing for all k in the range $[0.5, 2.0]$?
Hint: Find the closed-loop characteristic polynomial and study stability for a discretized set of k values.

Problem 3.9. Consider the plant in Example 3.1. Find the functions $T_1(s)$ and $T_2(s)$ associated with Youla parmeterization for this plant. Let $\tilde{Q}_Y = K$, use root-loci to explore the migration of zeros of the sensitivity functions $S(s)$ as K varies. Select a value of K that places a zero as close to the $j\omega$-axis as possible.

Problem 3.10. Write an M-file function (See, for example, Section A.3 of Appendix A) that takes as inputs the polynomials

$$n_Q(s), d_Q(s), n_P(s), d_+(s), \text{ and } d_-(s)$$

and produces as an output the compensator $C(s)$ [See (3.14)]. Note that the unstable numerator polynomial $d_+(s)$ divides the denominator polynomial in (3.14).
Hint: Use the MATLAB functions conv(a,b) to multiply two polynomials $a(s)$ and $b(s)$, and the function deconv(a,b) to divide the polynomial $a(s)$ by the polynomial $b(s)$.

Chapter 4

Digital Control Design

4.1 Introduction

Figure 4.1 shows a typical feedback digital control structure. The design objective is to find a digital compensator $D(z)$ that, with analog-to-digital (A/D) and digital-to-analog (D/A) converters and an analog plant, will result in a closed-loop system that is stable and meets given design specifications.

If we assume that the D/A converter is a zero-order-hold (ZOH) circuit, with sampling time T, then the analog plant $P(s)$ can be replaced by its equivalent *pulse tranfer function* $G(z)$ given by

$$G(z) = (1 - z^{-1})\mathcal{Z}\{\frac{P(s)}{s}\} \qquad (4.1)$$

where $\mathcal{Z}\{X(s)\}$ denotes the Z-transform of the sequence generated from the inverse Laplace transform of $X(s)$ sampled at time T; and z denotes the Z-transform variable. The details for the computation of $G(z)$ may be found in standard texts on control, e.g., [18] or [38]. It is

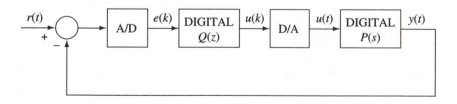

Figure 4.1: Digital Control Structure

also possible to use the MATLAB function *c2d(sys, Ts)* to compute $G(z)$ numerically, given the analog transfer function $P(s)$, specified by *sys*, and the sampling time *Ts*. The problem is then reduced to a feedback problem as shown in Figure 1.1, with the analog transfer functions $C(s)$ and $P(s)$ replaced by the digital transfer functions $D(z)$ and $G(z)$, respectively. In this chapter we will show how the interpolation theory developed in Chapters 2 and 3 can be modified to accomplish *direct digital design* of the compensator $D(z)$.

It is also possible to *design by emulation*, see, for example, Section 8.3 of [18]. For design-by-emulation, one can compute an analog compensator $C(s)$ using the theory in Chapters 2 and 3 and then with a mapping, $s = f(z)$, obtain the digital compensator $D(z)$. A common mapping used for this purpose is the so-called Tustin map

$$s = \frac{2}{T} \frac{z-1}{z+1} \tag{4.2}$$

The digital compensator is then given by

$$D(z) = C(s)|_{s=f(z)} \tag{4.3}$$

With design-by-emulation, the theory in this chapter is not needed, but some simulation is required to guarantee that the continuous-time design specifications have been met. The digital compensator $D(z)$, for a Tustin map, may be computed from the MATLAB function *c2d(sys, Ts, 'tustin')*.

4.2   Internal Stability

All the interpolation theory techniques developed in Chapters 2 and 3 may then be applied to designing compensators $D(z)$, with only a modification of the conditions for BIBO stability. In particular for discrete-time systems, a transfer function test for BIBO stability is given by the following:

A discrete-time system with rational transfer function $T(z)$ is discrete-time bounded-input-bounded-output (DT-BIBO) stable if and only if the transfer function is proper and all its poles are inside the unit circle.

Actually for discrete-time systems, bounded-inputs may produce bounded-outputs even if $T(z)$ is not *proper*, however improper rational functions correspond to *noncausal* systems and for real-time control all transfer functions must be causal, so the properness condition is included as part of the stability definition. As in the definition for analog systems, a discrete-time feedback system is internally stable if the following three transfer functions are DT-BIBO stable

$$E(z)/R(z) \;=\; \frac{1}{1 + D(z)G(z)} \tag{4.4}$$

$$Y(z)/M(z) \;=\; \frac{G(z)}{1 + D(z)G(z)} \tag{4.5}$$

$$U(z)/R(z) \;=\; \frac{D(z)}{1 + D(z)G(z)} \tag{4.6}$$

where $M(z)$ is the Z-transform of a discrete-time disturbance signal $m(k)$ added to the input $u(k)$; and $R(z)$ and $Y(z)$ are the Z-transforms of the signals $r(t)$ and $y(t)$ sampled at the sampling rate T (See Figure 4.1).

4.3 Deadbeat Response

A stable discrete-time (DT) system with transfer function $T(z)$ may have all its poles at the origin so that its transfer function takes on the form

$$T(z) = b_0 + b_1 z^{-1} + b_2 z^{-2} + ... + b_n z^{-n} \tag{4.7}$$

A transfer function of this type is also called a finite-impulse-response (FIR) filter. Such a system has the property that the step response reaches a steady-state value in a *finite* number of steps. This type of response is peculiar to DT systems and is called *deadbeat* response. Digital systems are often designed to have a deadbeat response.

In [29] the concept of "internal stability" is extended to DT systems where all of the transfer functions in (4.4)–(4.6) are "deadbeat." In [29], a DT system that is internally stable in this sense is referred to as a *Finite Input-Sequence: Finite Output-Sequence (FIFO)* system.

4.4   Digital Unit-Parameter Design

BIBO stability for DT systems requires that all the transfer-function poles must be inside the unit circle in the z-domain. This leaves a "hole" on the real axis of the z-plane that requires some attention in applying the p.i.p. condition to DT systems. For discrete-time systems we must look for interlacing of poles and zeros as we move from the $+1$ point to $+\infty$ and then move from $-\infty$ to -1 point. For example, the function

$$G(z) = \frac{(z-3)(z+1)}{(z-2)(z+4)}$$

does not satisfy the DT p.i.p. because in moving as indicated here, there is an odd number of poles, i.e., the single pole at $z = -4$, between the zeros at $z = 3$ and $z = -1$. On the other hand the function

$$G(z) = \frac{(z-2)(z+1)}{(z-3)(z+4)}$$

does satisfy the DT p.i.p. since "between" the zeros at $z = 2$ and $z = -1$ there are an even number of poles, i.e. the two poles at $z = 3$ and $z = -4$.

As in the analog case [with the analog unit $U(s)$ replaced by the DT unit $W(z)$], the compensator

$$D(z) = \frac{W(z) - D_g(z)}{N_g(z)} \tag{4.8}$$

where

$$G(z) = \frac{N_g(z)}{D_g(z)} \tag{4.9}$$

with $N_g(z)$ and $D_g(z)$ DT-BIBO stable functions, is a stable stabilizing compensator if $W(z)$ is a DT-BIBO unit that interpolates to

$$W(b_i) = D_g(b_i) \tag{4.10}$$

where b_i are the zeros of $N_g(z)$ *outside the unit circle*. A function $W(z)$ is a DT-BIBO unit if it is DT-BIBO and its inverse is also DT-BIBO. Recall that for DT-BIBO stability we require that the denominator polynomial of $N_g(z)$ and $D_g(z)$ to have all its zeros inside the unit circle. A polynomial with these properties will be referred to as a

Schur polynomial. Unfortunately, conditions for a polynomial to be Schur are more complicated than the conditions for a polynomial to be Hurwitz. For example, the second-order polynomial

$$z^2 + a_1 z + a_2$$

is Schur if and only if

$$|a_2| < 1 \quad \text{and} \quad |1 + a_2| < |a_1|$$

Example 4.1.
Consider the problem of stabilizing the plant

$$G(z) = \frac{z - 1.5}{z(z - 1)}$$

with a stable compensator.

Solution
Since there are no poles between the zeros at $z = 1.5$ and $+\infty$, the DT p.i.p. is satisfied, and hence a stable stabilizing compensator does exist. Let

$$N_g(z) = \frac{z - 1.5}{z^2} \quad \text{and} \quad D_g(z) = \frac{z - 1}{z}$$

The interpolation conditions on $W(z)$ are

$$W(1.5) = D_g(1.5) = 1/3 \quad \text{and} \quad W(\infty) = D_g(\infty) = 1$$

A first-order $W(z)$ that meets the condition at ∞ may be written

$$W(z) = \frac{z + a}{z + b}$$

where a and b satisfy, $|a| < 1$ and $|b| < 1$. The interpolation condition at $z = 1.5$ yields the equation

$$b = 3 + 3a$$

If we select $a = -0.8$, then $b = 0.6$ and an acceptable interpolating unit is given by

$$W(z) = \frac{z - 0.8}{z + 0.6}$$

If this unit is substituted back into Equation (4.8), the following stable stabilizing compensator results

$$D(z) = -\frac{0.4z}{z + 0.6}$$

4.5   Digital Q-Parameter Design

The procedure for Q-parameter design in the discrete-time domain is exactly the same as in the continuous-time domain discussed in Chapter 3. In particular, the DT compensator is given by

$$D(z) = \frac{Q(z)}{1 - Q(z)G(z)} \qquad (4.11)$$

where

$$Q(z) = B(z)\tilde{Q}(z) \qquad (4.12)$$

where $B(z)$ is any DT-BIBO function that is zero at the poles of $G(z)$ outside the unit circle and $\tilde{Q}(z)$ is any DT-BIBO stable function that satisfies the interpolations conditions

$$\tilde{Q}(a_i) = \frac{1}{\tilde{G}(a_i)} \qquad (4.13)$$

where

$$\tilde{G}(z) = B(z)G(z) \qquad (4.14)$$

To simplify the discussion, we have assumed that the unstable plant poles (poles outside the unit circle) are simple.

The sensitivity function is given by

$$E(z)/R(z) = S(z) = \frac{1}{1 + D(z)G(z)} = (1 - Q(z)G(z)) \qquad (4.15)$$

and the steady-state error to a step in reference signal is given by

$$e_{ss} = 1 - Q(1)G(1) \qquad (4.16)$$

Equation (4.16) follows directly from the DT final value theorem

$$e_{ss} = (1 - z^{-1})E(z)|_{z=1}$$

The poles of $S(z)$ may be selected by fixing the poles of $Q(z)$ and picking zeros of $Q(z)$ to cancel undesired poles of $G(z)$.

Example 4.2. Consider the plant of Example 4.1. Find a compensator, not necessarily stable, which stabilizes this plant and has a deadbeat error response to a step-input reference signal.

Solution. In order to obtain a deadbeat response we select

$$B(z) = \frac{z - 1}{z}$$

Then

$$\tilde{G}(z) = \frac{z - 1.5}{z^2}$$

and

$$\tilde{Q}(1) = \frac{1}{\tilde{G}(1)} = -2$$

Note that since $G(z)$ has a pole at $z = 1$, the steady-state error is automatically guaranteed to be zero. The function $Q(z)$ is then

$$Q(z) = B(z)\tilde{Q}(z) = -2\frac{z - 1}{z}$$

If this $Q(z)$ is substituted back into Equation (4.11), we obtain the compensator

$$D(z) = -\frac{2z}{z + 3}$$

The error response for the closed-loop system is given by

$$e(k) = v(k) + 2v(k - 1) - 3v(k - 2)$$

where here $v(k)$ is used to denote the DT unit step function, i.e.,

$$v(k) = 1, \ k \geq 0, \ \text{and} \ v(k) = 0, \ k < 0$$

It is obvious from this expression for $e(k)$ that the error is reduced to zero in two sampling intervals. Note that the compensator is unstable, with a pole at $z = 3$.

4.6   Notes and References

One of the first books on the design of digital control systems is the text of Ragazzini and Franklin [39], published in 1958. In this text the term *prototype response* is used instead of *deadbeat response*. As noted in Chapter 1, this text is the first to include a discussion of the interpolation approach to feedback system design (See in particular Section 7.5). A more recent book that deals with the analytical design of discrete-time systems is the text of Kučera [28]. In [28] Q-parameter concepts are introduced, from the polynomial point of view. In this approach the Bezout identity is used to obtain the Q-parameterization, rather than interpolation as we have done here.

4.7   Problems

Problem 4.1. Which of the following plants can be stabilized with a *stable* compensator?

$$G_1(z) = \frac{(z-3)(z+1)}{(z-2)(z+4)}$$

$$G_2(z) = \frac{(z-3)(z+1)}{(z-1.5)(z-2)}$$

$$G_3(z) = \frac{(z-1)(z-2)}{z(z-3)}$$

$$G_4(z) = \frac{1}{(z-1)^2}$$

Problem 4.2. Find a stable stabilizing compensator for the plants in problem 1, when one exists.

Problem 4.3. For the plants in problem 4.1 that cannot be stabilized with a stable compensator, find an unstable compensator that will stabilize the plant.

Problem 4.4. Consider an analog plant with transfer function

$$P(s) = \frac{1}{s-1}$$

Assume a ZOH D/A converter and a sampling time of $T = 1$, design a digital compensator $D(z)$ such that the closed-loop system has a zero steady-state deadbeat response to a unit-step reference input.

Problem 4.5. Use root-locus analysis to explore the possibility of stabilizing any of the plants in problem 4.1 by simple proportional feedback.

Problem 4.6. Consider the analog plant corresponding to an inverted pendulum with delay, with transfer function $P(s)$ given by (1.16). Let $g/L = 4$, $L/M = 10$, and $\tau = 2$. Then $P(s)$ becomes

$$P(s) = 10\frac{(1-s)}{(s+2)(s-2)(1+s)}$$

(a) If the sampling time is $T = 0.1$, use the MATLAB function *c2d* to compute the pulse tranfer function $G(z)$ for this plant.
(b) Use digital Q-parameter theory to design a digital compensator $D(z)$ that places all the closed-loop poles of the sensitivity function $S(z)$ inside a circle of radius 0.6.
(c) Can $G(z)$ be stabilized with a *stable* $D(z)$?

Problem 4.7. Consider the analog plant

$$P(s) = \frac{s-2}{(s-1)(s+2)}$$

(a) Find a *stable* digital compensator $D(z)$ that stabilizes this plant if $T = 0.1$.
(a) Repeat (a), if possible, for $T = 1.0$.

Problem 4.8. Consider the benchmark problem discussed in Example 3.3 with analog plant

$$P(s) = \frac{2}{s^2(s^2+4)}$$

Use digital Q-parameter theory to design a digital compensator $D(z)$ that places all the poles sensitivity function $S(z)$ at $z = 0.1$, given a sampling time $T = 0.5$.

Chapter 5

Robust Design

5.1 Introduction

In all of the previous chapters we have assumed that the plant to be controlled was perfectly known. In practice this is never the case. There is always some uncertainty in the plant, due to unforseen variations in operating conditions, incomplete knowledge of plant parameter values, etc. The design of systems that operate satisfactorily in the presence of plant uncertainties is called *robust design*. Feedback, properly designed, is the perfect structure for robustness. Indeed the original feedback patent by Black in 1927, [3], was precisely for this purpose.

The analytic design procedures developed in the previous chapters have a degree of robustness, due to the fact that the roots of a polynomial depend *continuously* on the coefficients of the polynomial. Thus, if a nominal closed-loop characteristic polynomial is stable, small enough perturbations on the coefficients, due, for example, to plant-parameter variations, will not make the polynomial go unstable. But "small enough" may not correspond to actual parameter variations. Therefore, it is essential to design systems that will tolerate *given levels of uncertainty*.

In this chapter we will pursue *robust analytic design* for four models of uncertainty. One is uncertainty due to pure gain variations in the plant. We refer to this as *gain margin design*. Gain margin design is discussed in most introductory textbooks on feedback control, e.g. [18]

and [38]. However, the design techniques are generally ad hoc, rather than analytic.

A second model we will discuss is commonly referred to as *multiplicative unstructured uncertainty.* We will focus on the analytic design of compensators that stabilize the closed-loop system in the presence of multiplicative unstructured uncertainty in the plant transfer function. The third model of uncertainty we will discuss is referred to as *multi-model uncertainty,* where one assumes that the plant can be a finite set of distinct plants. Stabilization of a finite set of possible plants with a single compensator is referred to as *simultaneous stabilization.* Our discussion, however, will be limited to the case of two plants, since this is the only case that can be solved analytically. The final model of uncertainty is associated with "passive" plants, that is, plants that can only absorb energy. For passive plants, strictly positive real (SPR) compensators can be used to guarantee robust stabilization of the closed-loop system.

5.2 Gain Margin Design

Let the plant transfer function be given by $kP(s)$, were k is an uncertain plant gain specified as a real number in the interval

$$k_1 \leq k \leq k_{max} \tag{5.1}$$

where k_1 is a fixed lower value of gain and k_{max} is an upper value. In general we wish to make k_{max} as large as possible, to have the largest possible *increasing gain margin.* We assume the nominal value of k is given by $k = 1$, and that $k_1 \leq 1$ Now the stability of a unity feedback system with compensator $C(s)$ and plant $kP(s)$ is determined by the zeros of the rational function

$$1 + kC(s)P(s) \tag{5.2}$$

since the zeros of this rational function are the poles of the closed-loop system. This function may be rewritten

$$(1 + k_1 C(s)P(s)) \left[1 + \frac{k - k_1}{k_1} \frac{k_1 C(s)P(s)}{1 + k_1 C(s)P(s)} \right] \tag{5.3}$$

For closed-loop stability we must guarantee that the square-bracketed term in (5.3) is stable and is not zero for all $Re\ s \geq 0$, and all k in the interval given in (5.1). This is equivalent to guaranteeing that

$$T(s) = \frac{k_1 C(s) P(s)}{1 + k_1 C(s) P(s)} \neq -\frac{k_1}{k - k_1} \tag{5.4}$$

for all $Re\ s \geq 0$ and $k_1 \leq k \leq k_{max}$. This, in turn, is equivalent to requiring that in the complex $T(s)$-plane, values along the negative real axis from $-\infty$ to $-\frac{k_1}{k_{max}-k_1}$, are not assumed when $Re\ s \geq 0$. Consider now the function

$$F(s) = \frac{k_1}{k_{max} - k_1} + \frac{k_1 C(s) P(s)}{1 + k_1 C(s) P(s)} \tag{5.5}$$

In the complex $F(s)$ the "forbidden" real line segment $[-\infty, -\frac{k_1}{k_{max}-k_1}]$ becomes the real line segment $[-\infty, 0]$. Therefore, to guarantee closed-loop stability we need to select a stable $F(s)$ such that for all $Re\ s \geq 0$

$$-\pi < arg\ F(s) < \pi \tag{5.6}$$

where $arg\ F(s)$ denotes the *argument* (phase) of the complex function $F(s)$. If $F(s)$ is selected with this property, then $F(s)$ can never assume real values in the interval $[-\infty, 0]$, as required for closed-loop stability. The gain margin design problem is then reduced to finding a stable function $F(s)$ that satisfies (5.6). However, to avoid "bad pole/zero cancellations" some interpolation conditions are induced on $F(s)$ at the unstable poles and zeros of the nominal plant $P(s)$. In particular, if b_i denotes unstable zeros of $P(s)$, including zeros at infinity, and a_i denotes unstable poles of $P(s)$, then to avoid cancellation with poles and zeros of the compensator tranfer function $C(s)$ we must have, from (5.5)

$$F(b_i) = \frac{k_1}{k_{max} - k_1} \tag{5.7}$$

and

$$F(a_i) = \frac{k_1}{k_{max} - k_1} + 1 = \frac{k_{max}}{k_{max} - k_1} \tag{5.8}$$

since $T(b_i) = 0$ and $T(a_i) = 0$. To simplify the interpolation problem we will assume that all the unstable poles and zeros are *simple*, i.e., have multiplicity one.

The gain margin design problem is therefore reduced to an interpolation problem with a function $F(s)$ that satisfies the argument condition in (5.6). But this interpolation problem can be reduced to an interpolation problem with a stable function $Z(s)$ with the argument constraint

$$-\pi/2 < Z(s) < \pi/2 \tag{5.9}$$

if we let $F(s) = [Z(s)]^2$, with the interpolation conditions

$$Z(b_i) = \sqrt{\frac{k_1}{k_{max} - k_1}}, \quad Z(a_i) = \sqrt{\frac{k_{max}}{k_{max} - k_1}} \tag{5.10}$$

A stable function $Z(s)$ that satisfies the argument condition (5.9) is called a *strictly positive real* (SPR) function.

The problem of gain margin design is then reduced to the problem of interpolation with an SPR function.

The interpolation points can be simplified further by first interpolating with $\tilde{Z}(s) = \sqrt{k_{max} - k_1} Z(s)$ with

$$\tilde{Z}(b_i) = \sqrt{k_1}, \quad \tilde{Z}(a_i) = \sqrt{k_{max}} \tag{5.11}$$

Note that $C(s)$ may be computed from (5.5) and expressed directly in terms of $\tilde{Z}(s)$, i.e.,

$$C(s) = \frac{[\tilde{Z}(s)]^2 - k_1}{k_1 P(s)(k_{max} - [\tilde{Z}(s)]^2)} \tag{5.12}$$

The general problem of interpolation with SPR functions is solved in reference [50]. In this chapter we will limit ourselves to interpolation problems that can be solved with first-order SPR functions, i.e., functions of the form

$$\tilde{Z}(s) = \frac{as + b}{s + c}, \quad a > 0, b > 0, c > 0 \tag{5.13}$$

The gain margin design problem then reduces to the following steps:

1. Find an SPR function $\tilde{Z}(s)$ that interpolates to the values given in (5.11), with as large a value of k_{max} as possible.

2. From $\tilde{Z}(s)$ and k_{max} determined in the previous step, compute $C(s)$ from (5.12).

It should be noted that the largest possible value of k_{max} that can be realized is limited by the fact that the interpolation conditions in (5.11) must be met by a *strictly positive real function*. This limitation will be illustrated in the following example. It also should be noted that if the plant is stable, the increasing gain margin can be made arbitrarily large. For stable plants, other design specifications, e.g., steady-state error, need to be added to the problem to limit k_{max}.

Example 5.1. Consider the plant

$$P(s) = \frac{s-1}{s-2}$$

and the fixed lower gain $k_1 = 1$. The problem is to design a compensator with a "large" increasing gain margin, i.e., a large value of k_{max}.

Solution. The interpolation conditions for this problem are given, from (5.11), by

$$\tilde{Z}(1) = 1, \quad \tilde{Z}(2) = \sqrt{k_{max}}$$

Assume a first-order SPR function of the form given in (5.13). The interpolation conditions then lead to the linear equations

$$a + b = 1 + c, \quad 2a + b = (2+c)\sqrt{k_{max}}$$

with solutions

$$a = (2+c)\sqrt{k_{max}} - (1+c), \quad b = 2(1+c) - (2+c)\sqrt{k_{max}}$$

The parameters a and b must all be positive for $\tilde{Z}(s)$ to be SPR. This leads to the following inequalities on k_{max}

$$\frac{(2+c)}{(1+c)} < \sqrt{k_{max}} < \frac{2(1+c)}{(2+c)}$$

The limiting value of k_{max}, of 2, is approached as $c \to \infty$. If the value $c = 10$ is selected then $\sqrt{k_{max}} = 1.8333$ or $k_{max} = 3.3611$. Also with this value of c we have $a = 10.9996$ and $b = 0.0004$, i.e., the first-order interpolating SPR function is given by

$$\tilde{Z}(s) = \frac{10.9996s + 0.0004}{s + 10}$$

From (5.12) the compensator is then given by

$$C(s) = -\frac{1.01939s + 0.84945}{s + 1.4286}$$

The Nyquist plot for the loop gain $C(s)P(s)$ is shown in Figure 5.1. Note that the loop gain has one unstable pole at $s = 2$, and that the Nyquist plot shows one counterclockwise encirclement of the -1 as required for closed-loop stability. Note also that the design gain margin is indeed achieved.

$\square$

5.3   Robust Stabilization

We assume now that the plant uncertainty may be characterized as follows

$$P(s) = P_0(s)[1 + m(s)] \tag{5.14}$$

where $P_0(s)$ denotes the nominal plant and $m(s)$ denotes a *multiplicative plant perturbation* term that represents plant uncertainty, but with known frequency bounds of the form

$$|m(j\omega)| \le |W_2(j\omega)|, \quad \text{for all } \omega \tag{5.15}$$

It is assumed that the perturbation $m(s)$ is stable, so that the nominal and perturbed plants have the same number of unstable poles. This model of plant uncertainty is commonly referred to as *unstructured multiplicative uncertainty*. See, for example, [12].

The analytic robust stabilization problem is then to find a compensator $C(s)$, when one exists, that preserves closed-loop stability for all plant perturbations $m(s)$ satisfying the bound given in (5.15). The given data for this problem is a nominal plant transfer function $P_0(s)$

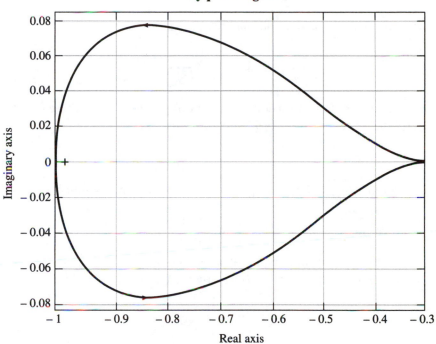

Figure 5.1: Nyquist Plot for Example 5.1

and an uncertainty bound function $W_2(s)$. The function $W_2(s)$ may be determined by computing bounds on

$$|m(j\omega)|^2 = \left|\frac{P(j\omega) - P_0(j\omega)}{P_0(j\omega}\right|^2 \tag{5.16}$$

For the analytic design procedure presented here, certain conditions are required for the bounding function $W_2(s)$. In particular, this function must be a BIBO unit, e.g., a BIBO stable function with a BIBO stable inverse.

Example 5.2. Consider the plant transfer function

$$P(s) = \frac{e^{-\tau s}}{s(s+1)}$$

where τ represents a time delay known only to lie in the range $0 \leq \tau \leq 0.1$. In this case the term in (5.16) becomes

$$|e^{-j\tau\omega} - 1|^2 = 2(1 - cos(\tau\omega))$$

As shown in Figure 4.1 of [17] a bound for this term is given by

$$W_2(s) = \frac{0.21s}{0.1s + 1}$$

This bound is not a unit, however it can be approximated by a unit by replacing the numerator term $0.21s$ by $0.21(s + 0.1)$.

$\square$

We show next that the robust stabilization problem can be reduced to a problem of interpolation with a strictly bounded real function (See Appendix A), or equivalently a strictly positive real function. The stability of a unity feedback system with compensator $C(s)$ and perturbed plant given by (5.14) is determined by the zeros of the function (all the zeros must be in the LHP)

$$1 + C(s)P_0(s)(1 + m(s)) \tag{5.17}$$

This function may be rewritten

$$(1 + C(s)P_0(s))[1 + m(s)T(s)], \tag{5.18}$$

where

$$T(s) = \frac{C(s)P_0(s)}{1 + C(s)P_0(s)} \tag{5.19}$$

Now with the assumption that the compensator $C(s)$ stabilizes the nominal system, the term $(1 + C(s)P_0(s))$ has all its zeros in the LHP, and if

$$|m(s)T(s)| < 1, \tag{5.20}$$

for all $Re \ s \geq 0$, then the square-bracketed term in (5.18) is never zero in RHP, guaranteeing that the function in (5.17) has zeros only in the LHP, i.e., that the closed-loop system is stable. Now since $m(s)$ and $T(s)$ are stable, and $|W_2(j\omega)|$ bounds $m(j\omega)$ for all ω, the maximum modulus theorem (See Appendix A) can be invoked to show that the inequality in (5.20) is satisfied if

$$|W_2(j\omega)T(j\omega))| < 1, \text{ for all } \omega \tag{5.21}$$

However for internal stability, "bad" pole-zero cancellations must be avoided. If a_i denotes the unstable poles of the nominal plant $P_0(s)$, and b_i the unstable zeros, this means that $T(s)$ must satisfy the interpolation conditions

$$T(a_i) = 1 \text{ and } T(b_i) = 0 \tag{5.22}$$

To simplfy the discussion it is assumed that all the unstable poles and zeros of the plant are simple. Let $B_z(s)$ be a stable all-pass function, i.e., $|B_z(j\omega)| = 1$, with all its zeros at the unstable zeros of $P_0(s)$, then if $T(s)$ satisfies the given interpolation conditions, the function

$$V(s) = \frac{W_2(s)T(s)}{B_z(s)} \tag{5.23}$$

is stable and satisfies the equation

$$|W_2(j\omega)T(j\omega)| = |V(j\omega)|, \text{ for all } \omega \tag{5.24}$$

The all-pass function $B_z(s)$ defined previously has the general form

$$B_z(s) = \frac{(s - b_1)(s - b_2)...}{(s + \bar{b}_1)(s + \bar{b}_2)...} \qquad (5.25)$$

where $\bar{b}_i$ denotes the complex conjugate of b_i

The robust stabilization problem then reduces to finding a stable function $V(s)$ that satisfies

$$|V(j\omega)| < 1, \quad \text{for all } \omega \qquad (5.26)$$

and interpolation conditions

$$V(a_i) = \frac{W_2(a_i)}{B_z(a_i)} \qquad (5.27)$$

A stable function $V(s)$ with magnitude bounded by one as in (5.26), is referred to as a *strictly bounded real* SBR function. The problem of interpolation with a SBR function is referred to as the *Nevanlinna-Pick* interpolation problem (See, for example, [12]). By the transformation

$$Z = \frac{1 + V}{1 - V} \qquad (5.28)$$

a SBR function V is mapped into a strictly positive real (SPR), and the algorithm for interpolation with SPR given in Appendix A can be used when there are many interpolation points.

The problem of robust stabilization is then reduced to the problem of interpolation with a SBR function.

For two interpolation points, say a_1 and a_2, one can always try a first-order SBR function of the form

$$V(s) = \frac{as + b}{s + c}$$

where $c > 0, |a| < 1$ and $|b| < c$.

Once the function $V(s)$ is computed, the compensator $C(s)$ may be computed from [see Equation (5.19)]

$$C(s) = \frac{T(s)}{P_0(s)(1 - T(s))} \tag{5.29}$$

where [See Equation (5.23)]

$$T(s) = \frac{B_z(s)V(s)}{W_2(s)} \tag{5.30}$$

It should be noted that for the compensator to be proper, $V(s)$ must have the same relative degree as the nominal plant $P_0(s)$. Also, $W_2(s)$ must be a BIBO unit for the same reason.

Example 5.3. This example is suggested by the example on page 192 of [17]. In particular, we select the following nominal plant and uncertainty bound

$$P_0(s) = \frac{s - 1}{(s + 1)(s - 0.5)}, \quad W_2(s) = 0.8\frac{(s + 0.1)}{(s + 1)}$$

The problem is to design a compensator that guarantees closed-loop stability for all multiplicative perturbations $m(s)$ that satisfy (5.15).

Solution. Given the unstable zero at $s = 1$, we select the all-pass function

$$B_z(s) = \frac{(s - 1)}{(s + 1)}$$

The interpolation condition, see (5.22), becomes

$$V(0.5) = \frac{W_2(0.5)}{B_z(0.5)} = -0.96$$

Since the relative degree of the nominal plant is one we try interpolating with

$$V(s) = \frac{a}{s + b}, \quad b > 0, \quad |a| < b$$

For large enough b this is a suitable SBR interpolation function. In particular, with $b = 13$ we obtain for the given interpolation point $a = -12.96$. The required SBR interpolating function is then given by

$$V(s) = \frac{-12.96}{s + 13}$$

From (5.30) the closed-loop transfer function is given by

$$T(s) = \frac{-16.2(s-1)}{(s+13)(s+0.1)}$$

and finally from (5.29) the compensator is given by

$$C(s) = \frac{-16.2(s+1)}{(s-29.8)}$$

5.4   Two-Plant Stabilization

The problem considered in this section is that of designing a single compensator $C(s)$ that simultaneously stabilizes two distinct plants $P_1(s)$ and $P_2(s)$. Problems of this type occur when a nonlinear system is linearized about two operating points and we wish to use the same compensator for both operating points, or when there are sensor or actuator failures that cause the plant dynamics to be radically altered.

In order to simplify the presentation we will assume that one of the plants, say P_1, is stable. Then from the Q-parameter theory developed in Chapter 3, we know that $P_1(s)$ is stabilized by the compensator

$$C(s) = \frac{Q(s)}{1 - P_1(s)Q(s)} \tag{5.31}$$

where $Q(s)$ is any *stable* function. If this same compensator is to stabilize $P_2(s)$, then the closed-loop characteristic polynomial, given by the numerator polynomial of

$$1 + C(s)P_2(s) = 1 + \frac{Q(s)P_2(s)}{1 - P_1(s)Q(s)} \tag{5.32}$$

must be a Hurwitz polynomial. But the numerator of (5.32) has the same zeros as

$$1 + (P_2(s) - P_1(s))Q(s) \tag{5.33}$$

Equation (5.33) is exactly the equation one would have for stabilization of the difference plant

$$P_\Delta(s) = P_2(s) - P_1(s) \tag{5.34}$$

by the stable "compensator" $Q(s)$. Thus, if one treats $Q(s)$ as a compensator, the U-parameter theory of Chapter 2 can be used to compute a stable stabilizing $Q(s)$, using the U-parameterization,

$$Q(s) = \frac{U(s) - D_\Delta(s)}{N_\Delta(s)} \qquad (5.35)$$

where $P_\Delta(s) = N_\Delta(s)/D_\Delta(s)$, with $N_\Delta(s)$ and $D_\Delta(s)$ stable rational functions. See the related discussion in Section 2.2. Recall that $U(s)$ is a unit that must interpolate to $D_\Delta(s)$, at the unstable zeros of $N_\Delta(s)$.

An immediate consequence of this result is the following existence theorem for the simultaneous stabilization of two plants: *Two plants $P_1(s)$ and $P_2(s)$, with $P_1(s)$ stable, can be stabilized by a single compensator $C(s)$ if and only if the "difference" plant $P_\Delta(s) = P_2(s) - P_1(s)$ satisfies the parity interlacing property (p.i.p.).*

Example 5.4. Consider the two plants

$$P_1(s) = \frac{2(2-s)}{(s+1)(s+2)}, \quad P_2(s) = \frac{(2-s)}{(s-1)(s+2)}$$

The problem is to find a single compensator that stabilizes both plants, if one exists.

Solution. In this case the difference plant is given by

$$P_\Delta(s) = \frac{(2-s)(3-s)}{(s-1)(s+1)(s+2)}$$

There are no poles between the zeros at $s = 2$ and $s = \infty$ so that p.i.p. is trivially satisfied. Hence, a simultaneously stabilizing compensator is known to exist. If one chooses, $h(s) = (s+1)^2(s+2)$, then

$$N_\Delta(s) = \frac{(2-s)(3-s)}{(s+1)^2(s+2)}, \quad D_\Delta(s) = \frac{(s-1)}{(s+1)}$$

The interpolation points for the unit $U(s)$ are given by

$$U(2) = D_\Delta(2) = 1/3, \quad U(3) = D_\Delta(3) = 1/2, \quad U(\infty) = D_\Delta(\infty) = 1$$

A first-order unit cannot interpolate these three points. Consider the second-order unit

$$U(s) = \frac{s^2 + as + b}{s^2 + cs + d}$$

The interpolation conditions lead to the linear equations

$$6a + 3b = -8 + 2c + d, \quad 6a + 2b = -9 + 3c + d$$

A study of the solutions for a and b indicates that both will be positive with $c = 1/2$ and $d = 11$. With this choice of c and d we have, $a = 5/12$ and $b = 1/2$, with the resulting unit

$$U(s) = \frac{12s^2 + 5s + 6}{12^2 + 6s + 132}$$

The function $Q(s)$ computed from (5.35) is then

$$Q(s) = \frac{23(s+2)(s+1)}{12s^2 + 6s + 132}$$

Finally, the compensator computed from (5.31) is

$$C(s) = \frac{23(s+2)(s+1)}{12s^2 + 52s + 40}$$

The closed-loop characteristic polynomials corresponding to P_1 and P_2 are, respectively

$$q_1(s) = 12s^2 + 52s + 40, \quad q_2(s) = 12s^3 + 17s^2 + 31s + 6$$

which can both be readily verified to be Hurwitz.

$\square$

5.5   Passive Systems

A plant with a transfer function $P(s)$ that satisfies the phase condition

$$-\pi/2 \le arg\, P(s) \le \pi/2, \quad \text{for all Re } s \ge 0 \tag{5.36}$$

is said to be a *passive* system (Also sometimes referred to as a *dissipative* system [22]). A function $P(s)$ that satisfies condition (5.36) is referred to as a *positive real* (PR) function. Note that a strictly positive real function, see (5.9), differs from a positive real function only in that the argument inequality is a *strict* inequality. The significance of passive systems is that they can be robustly closed-loop stabilized in a simple way, i.e., by selecting strictly positive real compensators $C(s)$. The reason for this follows from the fact that if $P(s)$ is PR and $C(s)$ is SPR, then

$$-\pi < arg\ C(s)P(s) < \pi, \text{ for all } Re\ s \geq 0 \qquad (5.37)$$

so that $1 + C(s)P(s)$ can never be zero in the RHP, i.e., the closed-loop poles must all be in the LHP.

Passive systems occur when the physical system has no internal sources of energy and the product of an input-output variable represents the derivative of energy, i.e., power. The impedance of an electrical circuit made up of capacitors, inductors, and resistors is an example of a PR function. Indeed, one of the first applications of PR functions was to analytic design (synthesis) of passive electrical networks (See, for example, the classical text of Guillemin [21]).

An example of a passive mechanical system is a simple mass-spring system, when the input is *force* on the mass and the output is *velocity.* In the time domain the equation for the mass/spring system is given by

$$M\ddot{x} = -kx + F \qquad (5.38)$$

where x is the diplacement of the mass, M is the mass, k is the spring constant, and F is the applied force. If the control input is taken to be $u = F$ and the output to be $y = \dot{x}$, then the product uy does represent mechanical power, and the transfer function

$$P(s) = V(s)/F(s) = \frac{s}{Ms^2 + k} \qquad (5.39)$$

It is possible to verify that this transfer function is PR for positive values of M and k, directly from a study of $arg\ P(s)$. However, the alternate conditions (See page 15 of reference [21]) given next are easier to test.

A function $Z(s)$ is PR if and only if the following three conditions are satisfied:

A. $Z(s)$ has no poles in the region, $Re\ s > 0$,
B. $Re Z(j\omega) \geq 0$ for all ω, and

C. poles on the $j\omega$-axis must be simple and must have positive partial-fraction coefficients.

For the mass-spring system we have:
1. Poles of $P(s)$ on the $j\omega$-axis $(s = \pm\sqrt{k/M})$,
2. $Re\, P(j\omega) = 0$, and
3. partial fraction expansion,

$$\frac{s}{Ms^2 + k} = \frac{1/2M}{s + j\sqrt{k/M}} + \frac{1/2M}{s - j\sqrt{k/M}}$$

From this we see that $P(s)$ is PR for *all* positive values of M and k.

Example 5.5. Consider the passive plant with transfer function $P(s)$ given in (5.39), and the SPR compensator

$$C(s) = \frac{1}{s+1}$$

Show that the closed-loop system for this $P(s)$ and $C(s)$ is indeed robustly stable for all positive M and k.

Solution. The closed-loop characteristic polynomial for this system, given by the numerator polynomial of $1 + C(s)P(s)$, is computed to be

$$p(s) = Ms^3 + Ms^2 + (k+1)s + k = a_0 s^3 + a_1 s^2 + a_2 s + a_3$$

Application of the Liénard-Chipart stability (See Section B.3 of Appendix B) test yields the following conditions for all roots of $p(s)$ to remain in the LHP

$$a_3 = k > 0,\, a_0 = a_1 = M > 0,\, a_1 a_2 - a_0 a_3 = M(k+1) - Mk = M > 0$$

These conditions are clearly met for all positive M and k.

□

Note that closed-loop stability is also preserved if the compensator is only PR, provided the plant is then SPR. This follows because the product $C(s)P(s)$ has the same properties demonstrated previously. This is useful if one wants to use compensators with a pole at the origin to eliminate steady-state errors. Robust stabilization with a PR compensator is illustrated in the next example.

Example 5.6. Consider two "wheels" with moments of inertia J_1 and J_2 connected with flexible shaft with stiffness coefficient k. Let the input to this plant be a control torque τ_1 applied at wheel 1, and let the output be the angular velocity of wheel 1, $\omega_1 = \dot{\theta}_1$. It is further assumed that wheel 2 has some rotational friction with frictional coefficient B. The equations for this system are given by

$$J_1\ddot{\theta}_1 = -k(\theta_1 - \theta_2) + \tau_1$$

and

$$J_2\ddot{\theta}_2 = -k(\theta_2 - \theta_1) - B\dot{\theta}_2$$

The transfer function relating the torque τ_1 to the angular velocity ω_1 is given by

$$P(s) = \frac{J_2 s^2 + Bs + k}{J_1 J_2 s^3 + J_1 B s^2 + k(J_1 + J_2)s + kB}$$

This transfer function will be SPR if $B > 0$ and PR if $B = 0$ We wish to study the closed-loop stability of this system when a PI compensator of the form

$$C(s) = K_P + \frac{K_I}{s}$$

is used for feedback control. Note that $C(s)$ given here is PR for all positive K_P and K_I. The problem is to show that for the particular PR compensator with $K_P = K_I = 1$, and plant with $J_1 = J_2 = k = 1$, the closed-loop system is robustly stable for any positive B.

Solution. The plant transfer function relating input τ_1 to output ω_1, computed for the given parameter values, is given by

$$P(s) = \frac{s^2 + Bs + 1}{s^3 + Bs^2 + 2s + B}$$

The closed-loop characteristic polynomial, computed from the numerator polynomial of the function $1 + C(s)P(s)$ is then given by

$$p(s) = s^4 + (B+1)s^3 + (3+B)s^2 + (2B+1)s + 1 = a_0 s^4 + a_1 s^3 + a_2 s^2 + a_3 s + a_4$$

For a fourth-order polynomial the Liénard-Chipart conditions become (See Section B.3 of Appendix B):

$$a_4 = 1 > 0, \ a_2 = (3 + B) > 0, \ a_0 = 1 > 0, \ \Delta_1 = a_1 = (B + 1) > 0$$

and

$$\Delta_3 = a_1 a_2 a_3 - a_1^2 a_4 - a_0 a_3^2 = 2B^3 + 4B^2 + 4B + 1 > 0$$

Clearly these inequalities are satisfied for any positive B.

$\square$

It should be noted that not all transfer functions associated with systems with no internal energy sources are SPR or PR. Sensors and actuators must be *collocated*, i.e., located at the same point, to insure passivity of the transfer function. For example, if the measured output (sensor) of the two-wheel example is the angle of wheel 2 ω_2, and the input (actuator) is the torque at wheel 1 τ_1, the transfer function relating τ_1 to ω_2 is not SPR or PR.

5.6   Notes and References

Modern robust control theory was developed largely in the period 1980–1990. A summary of robust results up to 1987 may be found in the reprint volume [10]. The Introduction in reference [10] contains an historical account of the development of robust control theory.

The use of interpolation methods for analytic gain margin design was first presented in 1980. See reference [43]. In the text of Doyle, Francis, and Tannenbaum [17], analytic design of gain and phase margins is discussed in some detail, however from the point of view of interpolation with *bounded real*, rather than *positive real* functions. Most introductory textbooks deal only with ad hoc design techniques for gain and phase margin design, and seldom consider plants that have both unstable poles and zeros. Here we have presented only the analytic design for gain margin. Analytic design for phase margin is a bit more involved and hence was omitted, but the theory is developed in [17] and further expanded in [15]. It is of interest to note, however, that the maximum phase margin, θ_{sup}, that can be achieved for a given plant is computable from an interpolation problem defined on the closed-loop transfer function $T(s)$. In particular, paraphrasing theorem 2 on page 202 of reference [17], we can say:

If P(s) is stable or minimum phase then $\theta_{sup} = \pi$, otherwise

$$\theta_{sup} = 2sin^{-1}\frac{1}{\gamma_{inf}} \tag{5.40}$$

where $T(s)$ interpolates as in (5.22), and γ_{inf} is the least upper bound on $|T(j\omega)|$, for all ω.

This result constitutes an important existence theorem for any approach to phase margin design.

In discussing robust stabilization for unstructured plant perturbations, we limited ourselves to *multiplicative* pertubations. Other uncertainty models are possible. For example, *additive* plant perturbations of the type

$$P(s) = P_0(s) + \Delta(s) \tag{5.41}$$

can be considered. Indeed, the early paper by Kimura [25] on robust stabilization for unstructured plant perturbations dealt with analytic design for this additive model.

Passive control is of special interest for flexible mechanical systems, where plant dynamics may be very complex. The monograph [22] details a number of applications of passive control for flexible mechanical systems. However, it is important to note that for a transfer function to be passive, input and ouput variables must be such that their product represents power at a given point. This implies that actuator and sensor must be collocated at the same point.

More details on the two-plant stabilization problem may be found in Vidyasagar [45], Section 5.4, or Dorato, Fortuna, and Muscato [12], Section 2.3. A complete treatment of the simultaneous stabilization problem may be found in Blondel [4].

When one attempts to realize a given analog compensator design, for example, with an operational amplifier circuit, a very important issue is, how precisely must the amplifier parameters be specified? A compensator where performance deteriorates significantly, perhaps even with a loss of stability, is said to be *fragile* [24]. This issue is the compensator counterpart of robustness with respect to plant uncertainties. In most of the designs presented in this chapter, compensator parameters were specified with accuracy up to five significant figures. How much can this "accuracy" be relaxed, without significantly deteriorating performance? We have not considered this issue here, but it does require consideration in the physical implementation of the designed compensator. It is expected that designs that provide robustness with respect to plant variations may require very accurate compensators, especially if the design is done for one specific performance measure. For example, an optimal design for increasing gain margin may require a very fragile compensator with respect to decreasing compensator gains.

Since gain and phase margin designs involve the product $C(s)P(s)$, these designs can be used for variations either in the *plant* or in the *compensator*. Thus, a gain margin design can provide a *nonfragile* design with repect to compensator gains, and a phase margin design can provide a *nonfragile* design with repect to pure time-delays in compensator implementation.

Finally, it should be noted that all the robust design procedures discussed in this chapter apply to the discrete-time systems of Chapter 4. The only modifications required are the replacement of the unit-circle as the stability boundary and the replacement of $s = j\omega$ by $z = e^{j\omega}$ for frequency response.

5.7   Problems

Problem 5.1. Consider the plant

$$P(s) = \frac{s+1}{(s-2)(s+2)}$$

(a) Show that for this plant the increasing-gain margin is unbounded. Design a compensator that guarantees an increasing-gain margin of $k_{max} = 10$

(b) Use Nyquist and Bode plots for this design to verify that the closed-

loop system is stable and that the design gain margin is realized.
(c) What is the *decreasing* gain margin for this design?

Problem 5.2. Consider the plant

$$P(s) = \frac{1}{s^2 - 1}$$

Repeat parts (a),(b), and (c) of problem 5.1 for this plant.
Note: This plant has a zero of multiplicity two at $s = \infty$. Our discussion was limited to problems with multiplicity one. If the zero at infinity is treated as a simple zero, the compensator may not be proper. Proceed with the computation of $C(s)$ as if the zero at infinity were simple.

Problem 5.3. Consider the delay perturbed plant described in Example 5.2, with nominal plant

$$P_0(s) = \frac{1}{s(s+1)}$$

and bounding function

$$W_2(s) = \frac{0.21(s + 0.1)}{0.1s + 1}$$

Find a compensator $C(s)$ that guarantees robust closed-loop stability.
Note: This nominal plant has relative degree two, therefore the bounding unit $V(s)$ must also have relative degree two.

Problem 5.4. Consider the uncertain plant

$$P(s) = \frac{1}{s - a}$$

where $0.5 \le a \le 1.5$.
Show that the unit

$$W_2(s) = \frac{0.01s + 0.5}{s + 0.5}$$

bounds the multiplicative uncertainty $m(s)$ [This can be done by computing $m(s)$ and verifying the bound (5.15)]. Use this bound and the nominal plant

$$P_0(s) = \frac{1}{s - 1}$$

to design a robustly stabilizing compensator $C(s)$.

Problem 5.5. Consider the two plants

$$P_1(s) = \frac{2}{s+p}, \quad P_2(s) = \frac{1}{s-p}$$

(a) Show that these two plants can be stabilized for any positive value of p.
(b) Find a simultaneously stabilizing compensator for these two plants when $p = 2$.

Problem 5.6. Consider the plant

$$P(s) = \frac{s-1}{s-a}, \quad a > 1$$

(a) Show that for this plant (See the Notes and Reference section)

$$\gamma_{inf} = \frac{a+1}{a-1} \tag{5.42}$$

Hint: Use

$$T(s) = \frac{(a+1)(s-1)}{(a-1)(s+1)}$$

to satisfy interpolations points for $T(s)$, then show that this implies

$$|T(j\omega)| = \frac{a+1}{a-1}$$

(b) Use (5.40) to compute the largest phase margin that can be designed for this plant.

Problem 5.7. Consider the "two-wheel" problem in Example 5.6.
(a) With the PI compensator

$$C(s) = 1 + \frac{1}{s}$$

and plant parameter values $J_1 = J_2 = B = 1$, show that the closed-loop system is stable for all positive values of k.
(b) Compute the transfer function, for the parameter values in part (a) and $k = 1$, relating the input torque τ_1 to the angular velocity of wheel two,

i.e., $w_2 = \dot{\theta}_2$. Show that this transfer function, with noncollacted sensor/actuator is *not* SPR.

Problem 5.8. Consider the benchmark problem in Example 3.3 in Chapter 3, with $M_1 = M_2 = k = 1$.
(a) Compute the transfer function relating force F to velocity $\dot{x}_1$, and verify that this transfer function is PR.
(b) Consider the SPR compensator

$$C(s) = 10\frac{s+1}{s+2}$$

Show that the closed-loop system is stable for any $k > 0$.
Hint: Use the Liénard-Chipart criterion.

Appendix A

Interpolation with SPR Functions

A.1 Preliminaries

The basic objective of this appendix is to develop an algorithm that can be used to compute a strictly positive real function $Z(s)$ that interpolates to given values z_i at points $s = a_i$, i.e., satisfies

$$Z(a_i) = z_i, \quad i = 1, ..., q \quad Re\ a_i > 0 \tag{A.1}$$

Our discussion here is limited to real rational functions of s, and to simplify the algebra the points a_i will be assumed to be real. The extension to complex points is straightforward and will be discussed subsequently. Some basic definitions first.

A function $Z(s)$ is said to be *strictly positive real (SPR)* if it is BIBO stable (all poles with negative real parts) and satisfies the argument condition

$$-\pi/2 < arg\ Z(s) < \pi/2 \tag{A.2}$$

for all $Re\ s \geq 0$.

A function $W(s)$ is said to be *strictly bounded real (SBR)* if it is BIBO stable and satisfies the magnitude condition

$$|W(s)| < 1 \tag{A.3}$$

for all $Re\ s \geq 0$.

The following link between SPR and SBR functions is critical in our discussion of interpolation with SPR functions:

The function $Z(s)$ is SPR if and only if the function

$$W(s) = \frac{Z(s) - z_1}{Z(s) + z_1} \tag{A.4}$$

is SBR, where z_1 is any positive real number.

Note that (A.2) is equivalent to

$$Re\ Z(s) > 0 \tag{A.5}$$

for all $Re\ s > 0$.

The proof that $Z(s)$ SPR implies that $W(s)$ is SBR, and conversely, follows by noting that for any complex numbers Z and W one has

$$1 - W\bar{W} = \frac{(2z_1)(Z + \bar{Z})}{|Z + z_1|^2} \tag{A.6}$$

Since $Re\ Z = 1/2(Z + \bar{Z})$ and $z_1 > 0$, it follows from (A.6) that $Re\ Z > 0$ if and only if $|W|^2 = W\bar{W} < 1$.

From complex variable theory, it is known that the argument and magnitude conditions given by (A.2) and (A.3), respectively, need be tested only for $s = j\omega$, when $Z(s)$ and $W(s)$ are BIBO stable functions. A proof of this result, referred to as the *maximum modulus theory* or *maximum modulus principle*, is beyond the scope of this introductory text. However, a proof can be found in most texts on complex variables. See, for example, Chapter 2 of reference [6].

A.2   Interpolation Algorithm

The SPR interpolation theory presented next follows the development found in reference [50]. We will show first that the problem of interpolating with a SPR function $Z_1(s)$, q points as in (A.1), can be reduced to interpolating with a SPR function $Z_2(s)$ with one less point. This iteration step is repeated until the problem is reduced to interpolation with a single point. A positive single point, say z_q, can always be interpolated with the constant SPR function $Z_q(s) = z_q$. The iterations are then reversed to compute $Z_1(s)$ from $Z_q(s)$.

The basic mapping that permits this reduction of interpolation points is given by

$$\frac{Z_1(s) - z_1}{Z_1(s) + z_1} = (\frac{s - a_1}{s + a_1})\frac{Z_2(s) - 1}{Z_2(s) + 1} \tag{A.7}$$

From (A.7) and (A.6) it follows that

1. If $Z_2(s)$ is SPR, then the right-hand side of (A.7) is SBR. Then from (A.4) it follows that $Z_1(s)$ must be SPR. Thus, in the direction Z_2 to Z_1, with $z_1 > 0$, we have an SPR to SPR mapping, with the single interpolation condition

$$Z_1(a_1) = z_1 \tag{A.8}$$

   holding for *arbitrary* $Z_2(s)$.

2. Conversely if $Z_1(s)$ is SPR and $z_1 > 0$, then along the $j\omega$-axis we have

$$\left| \frac{Z_2(j\omega) - 1}{Z_2(j\omega) + 1} \right| < 1 \tag{A.9}$$

   since

$$\left| \frac{j\omega - a_1}{j\omega + a_1} \right| \text{ and } \left| \frac{Z_2(j\omega) - 1}{Z_2(j\omega) + 1} \right| < 1 \tag{A.10}$$

   But (A.9) implies that $Re\, Z_2(j\omega) > 0$.

   Now from the basic mapping (A.7) one may write

$$\frac{Z_2(s) - 1}{Z_2(s) + 1} = (\frac{s + a_1}{s - a_1})\frac{Z_1(s) - z_1}{Z_1(s) + z_1} \tag{A.11}$$

   Note that the unstable pole, at $s = a_1$, which appears in the right-hand side of (A.11) is actually canceled since $Z_1(s)$ interpolates to z_1 at $s = a_1$. Thus, we have that the inverse mapping given by (A.11) is an SPR to SPR mapping from $Z_1(s)$ to $Z_2(s)$, as long as $Z_1(a_1) = z_1$.

To illustrate how the interpolation algorithm works, consider the case of three points, i.e., $q = 3$. The algorithm can best be explained by the array shown next, which we will refer to as the *SPR array* :

$$
\begin{array}{llll}
a_1 & a_2 & a_3 & \leftarrow a_i \\
z_1 & z_2 & z_3 & \leftarrow Z_1(s) \\
 & z_2^{(1)} & z_3^{(1)} & \leftarrow Z_2(s) \\
 & & z_3^{(2)} & \leftarrow Z_3(s)
\end{array}
$$

The notation $z_i^{(1)}$ is used to denote the first-iteration values (in general $q - 1$ values), and $z_i^{(2)}$ to denote the second-iteration values ($q - 2$ values), etc. The first-iteration values $z_2^{(1)}$ and $z_3^{(1)}$ are computed from the basic mapping by letting $Z_1(a_2) = z_2$ and solving for $Z_2(a_2) = z_2^{(1)}$ with $s = a_2$, and by letting $Z_1(a_3) = z_3$ and solving for $Z_2(a_3) = z_3^{(1)}$ with $s = 3$. Note that in the first iteration only two points need be interpolated since for any $Z_2(s)$ we have $Z_1(a_1) = z_1$. The last row is the final row in this case, i.e., $q = 3$ and $z_3^{(2)}$ is computed from the basic mapping moved to the second point a_2, i.e., the mapping

$$
\frac{Z_2(s) - z_2^{(1)}}{Z_2(s) + z_2^{(1)}} = (\frac{s - a_2}{s + a_2}) \frac{Z_3(s) - 1}{Z_3(s) + 1} \tag{A.12}
$$

That is, $z_3^{(2)}$ is computed by letting $Z_2(a_3) = z_3^{(1)}$ and solving for $Z_3(a_3) = z_3^{(2)}$. The last value $z_3^{(2)}$ may be interpolated by the constant SPR function

$$
Z_3(s) = z_3^{(2)} \tag{A.13}
$$

The function $Z_1(s)$, which is now guaranteed to interpolate the three values (z_1, z_2, z_3) is then computed by first computing $Z_2(s)$ from $Z_3(s)$ using the second-iteration mapping (A.12), and then computing $Z_1(s)$ from $Z_2(s)$ using the initial mapping (A.7). This backward iteration will obviously produce a properly interpolating SPR function $Z_1(s)$ if each of the "diagonal" values z_1, $z_2^{(1)}$, and $z_3^{(2)}$ are *positive*. It can be shown that this condition is also necessary for the existence of an interpolating SPR function. We have therefore :

A necessary and sufficient condition for the existence of an interpolating SPR function is that the "diagonal" elements of the SPR array must all be positive.

If an interpolation point is at infinity, this point should be listed last in the SPR array, since the basic mapping, required to move down a row, breaks down in this case. The same is true for an interpolation point at $s = 0$. It should be listed last. If there are interpolation points at both infinity and zero, then they both should be moved to the last two places in the row of interpolation points. The last row of the SPR interpolation array, which consists now of two interpolation values, one for $s = \infty$ and the other for $s = 0$, can always be interpolated by the first-order SPR function

$$Z(s) = \frac{as + b}{s + 1}, \ a > 0, \ b > 0 \tag{A.14}$$

as long as the two interpolation values are positive.

If complex points are to be interpolated, they must appear in complex conjugate pairs if the function $Z_1(s)$ is to be *real*. For complex points, the basic mapping must be modified as follows:

$$\frac{Z_1(s) - z_1}{Z_1(s) + \bar{z}_1} = (\frac{s - a_1}{s + \bar{a}_1}) \frac{Z_2(s) - 1}{Z_2(s) + 1} \tag{A.15}$$

where $\bar{a}_1$ and $\bar{z}_1$ denote the complex conjugates of a_1 and z_1, repectively, and $Re \ a_1 > 0$, $Re \ z_1 > 0$. If there are interpolation points on the $j\omega$-axis, $Re \ a_1 = 0$, and this mapping breaks down. Some modification is required in this case . A suitable modification for interpolation on the $j\omega$-axis is given in reference [42].

Listed next is an M-file for a MATLAB function *spr(pts,val)* which will compute an interpolating SPR function for an arbitrary number of real interpolating points.

M-file for Computation of Interpolating SPR Functions

```
function spr(pts, val)
%This M file spr.m computes an interpolating SPR function
%given the interpolation points (pts) and interpolation
%values (val). The diagonal values of the SPR interpolation
%array are displayed to check for positivity.
%Computation of diagonal values:
q=length(pts);
for k=2:q
for j=k:q
num(j)=(val(j)+val(k-1))*(pts(j)-pts(k-1))+ ...
(val(j)-val(k-1))*(pts(j)+pts(k-1));
den(j)=(val(j)+val(k-1))*(pts(j)-pts(k-1))- ...
(val(j)-val(k-1))*(pts(j)+pts(k-1));
val(j)=num(j)/den(j);
end
end
disp('diagonal values of SPR array:')
disp(val)
%Computation of interpolating SPR function, Z=numZ/denZ:
numZ=val(q);
denZ=1;
for i=q-1:-1:1
sum=numZ+denZ;
dif=numZ-denZ;
numZ=val(i)*[conv([1 pts(i)],sum)+conv([1 -pts(i)],dif)];
denZ=[conv([1 pts(i)],sum)-conv([1 -pts(i)],dif)];
end
disp('SPR numerator polynomial vector:')
disp(numZ)
disp('SPR denominator polynomial vector:')
disp(denZ)
```

Example A.1. Consider the problem of maximizing the increasing gain margin for the plant

$$P(s) = \frac{(s-2)(s-3)}{(s-1)(s+3)}$$

given $k_1 = 1$. Using the theory developed in Chapter 5, we know that this problem reduces to finding a SPR function $Z_1(s) = \tilde{Z}(s)$ that interpolates to the values

$$Z_1(1) = \sqrt{k_{max}}, \quad Z_1(2) = 1, \quad Z_1(3) = 1$$

Solution. In this case the interpolation points are $a_1 = 1, a_2 = 2$, and $a_3 = 3$, and the interpolation values are $z_1 = \sqrt{k_{max}}, z_2 = 1$, and $z_3 = 1$. The SPR array for this problem becomes:

$$
\begin{array}{ccc}
1 & 2 & 3 \quad \leftarrow a_i \\
z_1 & 1 & 1 \quad \leftarrow Z_1(s) \\
 & z_2^{(1)} & z_3^{(1)} \quad \leftarrow Z_2(s) \\
 & & z_3^{(2)} \quad \leftarrow Z_3(s)
\end{array}
$$

From the mappings given in (A.7) and (A.12), the following values may be computed

$$z_2^{(1)} = \frac{-2z_1 + 4}{4z_1 - 2}, \quad z_3^{(1)} = \frac{3 - z_1}{3z_1 - 1}$$

$$z_3^{(2)} = \frac{-1 + 1.4z_1}{z_1(1.4 - z_1)}$$

For an interpolating SPR function to exist, all of these "diagonal" values $(z_1, z_2^{(1)}, \text{ and } z_3^{(2)})$ must be positive. A study of these values indicates that we must have

$$z_1 = \sqrt{k_{max}} < 1.4$$

If we select $z_1 = 1.3$ (This corresponds to $k_{max} = 1.69$), the interpolation points are given by the vector $pts = [1\ 2\ 3]$ and the interpolation values by the vector $val = [1.3\ 1\ 1]$. With this data the M-file function $spr(pts,val)$ listed previously yields the SPR interpolation function

$$Z_1(s) = \frac{14.35s^2 + 9.75s + 65.6}{4s^2 + 61.5s + 3.5}$$

A.3   Interpolating Units via SPR Functions

As shown in reference [13], biproper (proper with proper inverse) SPR functions can be used to compute interpolating units. Consider the problem of finding a unit $U(s)$ that interpolates at the points a_i, i.e.,

$$U(a_i) = \mu_i, \quad i = 1, ..., q \tag{A.16}$$

Again, to simplify the development we will assume the points a_i are all real. There is no loss of generality in assuming the points are all positive when the a_i are real, since otherwise one could interpolate with $-U(s)$. Consider the m-th root of each value μ_i, i.e.,

$$z_i = (\mu_i)^{\frac{1}{m}} \tag{A.17}$$

As m approaches infinity all the values z_i approach one. But when all the values cluster close enough to one, an interpolating SPR function always exists (perhaps not too obvious, but clear in the limiting case when all the values are exactly equal to one). The p.i.p. conditon on the plant $P(s)$ insures that all the interpolation values μ_i have the same sign, which is all that is required for an interpolating unit to exist. Without loss of generality we will assume that the values are all positive.

Thus, to interpolate values μ_i with a unit $U(s)$, one can first interpolate the values $z_i = (\mu_i)^{\frac{1}{m}}$ (with m selected sufficiently large) with an SPR function $Z(s)$ and then compute the required unit from

$$U(s) = [Z(s)]^m \tag{A.18}$$

This works because a product of SPR functions is always a unit. For more details see [13].

Example A.2. Consider the plant of problem 2.3, i.e.,

$$P(s) = \frac{(s-1)(s-2)}{(s-0.9)(s+1)}$$

This plant satisfies p.i.p., hence a stable stabilizing compensator exists. If we select $h(s) = (s+1)^2$,

$$D_p(s) = \frac{(s-0.9)}{(s+1)}$$

the interpolation points for the unit $U(s)$ required to design a stable compensator are given by

$$U(1) = D_p(1) = \mu_1 = 1/20, \quad U(2) = D_p(2) = \mu_2 = 11/30$$

The problem is to find a unit that interpolates to these points.

Solution. For $m = 1$ the SPR array for this problem is given by (recall that for $m = 1$, $z_i = \mu_i$):

$$
\begin{array}{lll}
1 & 2 & \leftarrow a_i \\
0.0500 & 0.3667 & \leftarrow Z_1(s) \\
 & -2.5390 & \leftarrow Z_2(s)
\end{array}
$$

Since not all the "diagonal" terms in the array are positive, $m = 1$ does not work, and at least a second-order unit is required for interpolation. For $m = 2$ the SPR array is given by (recall that for $m = 2$, $z_i = (\mu_i)^{1/2}$):

$$
\begin{array}{lll}
1 & 2 & \leftarrow a_i \\
0.2236 & 0.6055 & \leftarrow Z_1(s) \\
 & -6.2383 & \leftarrow Z_2(s)
\end{array}
$$

Again, not all the "diagonal" terms are positive, so that $m = 2$ does not work, and at least a third-order unit is required. For $m = 3$ the SPR array is given by (recall that for $m = 3$, $z_i = (\mu_i)^{1/3}$):

$$
\begin{array}{lll}
1 & 2 & \leftarrow a_i \\
0.3684 & 0.7157 & \leftarrow Z_1(s) \\
 & 50.2820 & \leftarrow Z_2(s)
\end{array}
$$

Now all the "diagonal" terms *are positive*, hence a third-order unit can interpolate the given points. Using the M-file listed previously for the function *spr(pts,val)* with

$$pts = [1\ 2], \quad val = [0.3684\ 0.7157]$$

one obtains the SPR interpolating function

$$Z_1(s) = \frac{37.1194s + 0.7368}{2s + 100.7583}$$

The required interpolating unit is then given by

$$U(s) = [Z_1(s)]^3$$

Appendix B

Stability Criteria

In the interpolation approach presented in the body of the text, we do not use classical stability criteria such as Nyquist, root-locus, and Routh-Hurwitz to design feedback systems. Indeed, except for simple proportional control, these classical results require trial-and-error procedures, and the purpose of this text supplement is to present analytic alternatives to trial-and-error design. However, we have used these classical tests to verify analytic results and to explore designs with simple "static" compensators, i.e., compensators of the form $C(s) = K$. So here we review briefly some stability criteria, e.g., Nyquist, root-locus, and Routh-Hurwitz, including software that is available for these criteria. It is assumed that the text being supplemented includes detailed discussions of these classical results, especially Nyquist and root-locus. Sections B.1 and B.2 contain brief reviews of Nyquist and root-locus. In Section B.3 we present the Liénard-Chipart stability criterion as an alternative to Routh-Hurwitz, since it requires less computation and is not covered in most introductory texts.

B.1 Nyquist Stability Criterion

The basic idea behind the Nyquist stability criterion is that the number of zeros of the function

$$1 + C(s)P(s) \tag{B.1}$$

in the RH s-plane is related to the number of encirclements of the -1 by the curve of $C(s)P(s)$ obtained as s varies along the Nyquist contour in the s-plane. The Nyquist contour is defined as the set of points in the s-plane going from $s = -j\infty$ to $s = j\infty$ and moving along an infinite semicircle that encloses the RHP. In particular (see, for example, Section 6.3 of [18]), if P denotes the number of poles of $C(s)P(s)$ in the RHP (referred to as open-loop poles), Z the number of zeros of $1 + C(s)P(s)$ in the RHP, and N the net number of clockwise encirclements of the -1 by $C(s)P(s)$ as s moves along the Nyquist contour, then we have the basic Nyquist result that

$$N = Z - P. \tag{B.2}$$

Since closed-loop stability requires that all the zeros of $1 + C(s)P(s)$ be in the LHP, Z should be zero and we have the stability criterion:

If $C(s)P(s)$ is stable there should be no net clockwise encirclements of the -1 point, and if $C(s)P(s)$ is unstable, with P unstable poles, there should be P counterclockwise encirclements of the -1 point.

If $C(s)P(s)$ has poles on the $j\omega$-axis, the Nyquist contour must be modified to avoid these poles. This is generally done by indenting the contour with "small" semicircles so that these poles are not inside the Nyquist contour, hence are not counted in P.

A plot of $C(j\omega)P(j\omega)$ in the complex plane is referred to as a *Nyquist plot*. Nyquist plots can be obtained from MATLAB software by using the MATLAB function *nyquist*. In particular, if $C(s)P(s)$ is expressed as a ratio of two polynomials $n(s)$ and $d(s)$, i.e.,

$$C(s)P(s) = \frac{n(s)}{d(s)} \tag{B.3}$$

and $n(s)$ and $d(s)$ are represented by MATLAB vectors *num* and *den* of coefficients (see [32] for details), then the MATLAB function *nyquist(sys)*, with *sys=tf(num.den)*, returns a Nyquist plot. A default set of frequency values is selected in this case for the Nyquist plot. If one wants to see a Nyquist plot with preselected frequency points, a MATLAB vector w of frequency must be added as a *nyquist* function argument, i.e., *nyquist(sys,w)*. This may be required when there are open-loop poles on the $j\omega$-axis to avoid poor default scaling of the Nyquist plot. The problem is compounded when all the poles and zeros are on the

$j\omega$-axis, for then the MATLAB Nyquist plot lies entirely on the real or entirely on the complex line and one gets essentially a "stealth" Nyquist plot. Also MATLAB software will not indicate what happens when "small" indentations are made to avoid poles on the $j\omega$-axis. The situation is improved a bit with a Bode plot (MATLAB function *bode*), but detecting closed-loop stability can still be difficult, as illustrated by Example B.2.

Example B.1. Consider the compensator

$$C(s) = 12\frac{(s+1)}{(s-7)}$$

which was designed to stabilize the plant

$$P(s) = \frac{s-1}{(s-2)(s+1)}$$

in Example 3.1 of text. The problem is to use a Nyquist plot to verify that the closed-loop system is indeed stable.

Solution. From the product

$$C(s)P(s) = \frac{(12s-12)}{(s^2 - 9s + 14)}$$

the following MATLAB steps produce the Nyquist plot shown in Figure 3.1.

```
>> num=[12 -12]
>> den=[1 -9 14]
>> sys=tf(num,den)
>> nyquist(sys)
```

Note that in this case the loop gain $C(s)P(s)$ has two poles in the RHP so that $P = 2$ and hence two counterclockwise encirclements are required of the -1 point. This is indeed the case shown in Figure 3.1.

Example B.2. Consider the plant

$$P(s) = -\frac{s(s^2+2)}{(s^2+1)(s^2+3)} = \frac{-s^3 - 2s}{s^4 + 4s^2 + 3}$$

Explore the use of a Nyquist plot obtained from the MATLAB function *nyquist* to determine closed-loop stability of the feedback system when $C(s) = 1$.

Solution. The plot shown in Figure B.1 is obtained from the follow MATLAB steps:

```
>> num=[-1 0 -2 0]
>> den=[1 0 4 0 3]
>> sys=tf(num,den)
>> nyquist(sys)
```

Note the large scale on the plot (due to poles on the $j\omega$-axis) and the "stealth" nature of the Nyquist plot. A Bode plot, obtained from the MATLAB function *bode(sys)* is shown in Figure B.2. However, even from the Bode plot, it is not easy to detect if the closed-loop system is stable. Actually the closed-loop characteristic polynomial in this case is given by

$$a(s) = s^4 - s^3 + 4s^2 - 2s + 3$$

Since some of the coefficients of this polynomial are negative, the closed-loop system is actually known to be unstable (See Section B3).

B.2   Root-Locus

Root-locus is the locus of points in the s-plane that satisfy the equation

$$1 + KP(s) = 0 \tag{B.4}$$

as K varies from zero to infinity. When simple proportional feedback, $C(s) = K$, is used for a plant with transfer function $P(s)$, the zeros of $1 + KP(s)$ are poles of the closed-loop system, hence the interest in the migration of zeros of $1 + KP(s)$ as K varies.

If $P(s)$ is expressed as a rational function

$$P(s) = \frac{n(s)}{d(s)} = \frac{b_0 s^m + b_1 s^{m-1} + \ldots + b_m}{s^n + a_1 s^{n-1} + \ldots a_n} \tag{B.5}$$

where $m \le n$, the rules for plotting the root-locus, see, for example Chapter 5 of [18], are normally stated for the case $Kb_0 > 0$. When

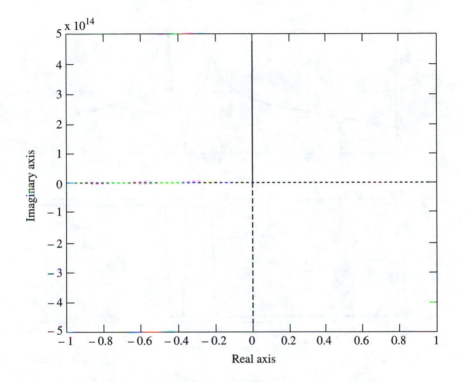

Figure B.1: Nyquist Plot for Example B.2

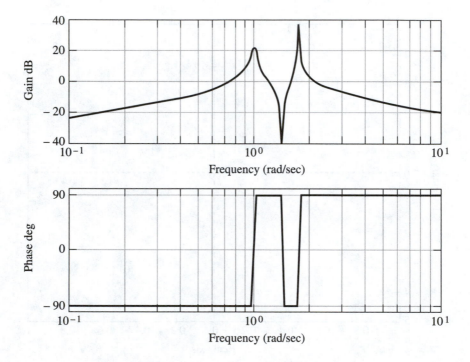

Figure B.2: Bode Plot for Example B.2

$Kb_0 < 0$, some of the rules change, and the the root-locus plot is referred to as a *complementary root-locus* plot. We will not repeat all the rules for plotting a root-locus here, but we will summarize a few rules that are especially useful. In particular:

1. As K increases from zero to infinity the roots migrate from the poles of $P(s)$ to the zeros of $P(s)$.

2. When $Kb_0 > 0$, a point on the real s-plane axis is part of the root-locus if to the right of the point there is an *odd* number of poles and zeros of $P(s)$, and when $Kb_0 < 0$, there is an *even* number of poles and zeros.

3. The root locus has $n - m$ (relative degree) equally spaced linear asymptotes.

These rules make it evident that plants, with transfer function $P(s)$, are difficult to control if the relative degree is three or greater, or there are zeros in the RHP. In both cases, too large a value of K will cause instability.

The MATLAB function *rlocus* may be used to obtain root-locus plots numerically. If $n(s)$ and $d(s)$ are expressed as coefficient vectors *num* and *den*, then the MATLAB function *rlocus(sys)*, with *sys=tf(num,den)*, returns a root-locus plot for $1 + KP(s)$, for a default set of positive values for K. One can obtain a complementary root-locus plot by replacing $n(s)$ by $-n(s)$. If one is not satisfied by the default selection of K-values, one can add an argument to the *rlocus* function, i.e., *rlocus(sys,k)* where k is a vector of selected values of K.

Example B.3. Consider the problem of exploring simple proportional feedback, i.e., $C(s) = K$, for the plant

$$P(s) = \frac{s - 2}{(s - 1)(s + 2)} = \frac{s - 1}{s^2 + s - 2}$$

discussed in Example 2.2, using the function *rlocus*.

Solution. The MATLAB steps

```
>> num=[1 -2]
>> den=[1 1 -2]
>> sys=tf(num,den)
>> rlocus(num,den)
```

produce the root-locus shown in Figure 2.1. If num=[1 -2] is replaced by num=[-1 2], the complimentary root-locus shown in Figure 2.2 is produced. Clearly this plant cannot be stabilized by simple proportional feedback since there are roots of $1 + KP(s)$ in the closed RHP for any value of K.

B.3   Liénard-Chipart Stability Criterion

A classic algebraic stability criterion, included in almost all introductory control texts is the *Routh-Hurwitz* (RH) criterion, sometimes referred to simply as the Routh stability criterion. This criterion gives conditions on the coefficients a_i of a polynomial,

$$p(s) = s^n + a_1 s^{n-1} + a_2 s^{n-2} + ... + a_n \tag{B.6}$$

in order that all its roots have *negative real parts*. A polynomial with these properties is commomly referred to as a *Hurwitz polynomial*. A simple *necessary* condition for the polynomial (B.6) to be Hurwitz is that all the coefficients a_i be *positive*. When $p(s)$ is the closed-loop characteristic polynomial, $p(s)$ Hurwitz is precisely the condition for closed-loop stability. The RH test is normally presented in terms of what is called the Routh array (See, for example, Section 4.4.3 of [18]). It can be shown that the RH criterion computed in terms of the Routh array is equivalent to the following conditions,

$$\Delta_1 > 0, \ \Delta_2 > 0, ..., \Delta_n > 0 \tag{B.7}$$

where Δ_i is the so-called Hurwitz determinant, given by

$$\Delta_i = det \begin{bmatrix} a_1 & a_3 & a_5 & \cdots & \\ 1 & a_2 & a_4 & \cdots & \\ 0 & a_1 & a_3 & \cdots & \\ 0 & 1 & a_2 & \cdots & \\ 0 & 0 & a_1 & \cdots & \\ \vdots & \vdots & \vdots & \vdots & \ddots \\ 0 & 0 & \cdots & \cdots & a_i \end{bmatrix}, \ a_k = 0, \text{ for } k > n \tag{B.8}$$

where $det[H]$ denotes the determinant of the matrix H.

The *Liénard-Chipart* (LH) criterion reduces the number of determinants that have to be evaluated. In particular, the LH criterion states that the polynomial (B.6) has all its roots with negative real parts *if and only if* any one of the following set of inequalities hold

$$a_n > 0, a_{n-2} > 0, ... : \Delta_1 > 0, \Delta_3 > 0, ... \tag{B.9}$$

or

$$a_n > 0, a_{n-2} > 0, ...; \Delta_2 > 0, \Delta_4 > 0, ... \tag{B.10}$$

While either of these sets of inequalities may be used, (B.9) is more efficient (fewer inequalities to test) for n *even*, and (B.10) is more efficient for n *odd*. Also note that the inequalities $\Delta_i > 0$ need be tested only up to $i = n - 1$.

A proof of the LH criterion is beyond the scope of an introductory course, but the details of a proof are given in reference [20] for the interested reader.

Example B.4. Consider the two polynomials

$$p_1(s) = s^3 + a_1 s^2 + a_2 s + a_3$$

and

$$p_2(s) = s^4 + a_1 s^3 + a_2 s^2 + a_3 s + a_4$$

The problem is to determine conditions on the coefficients for the two polynomials to be Hurwitz.

Solution. For the polynomial $p_1(s)$, $n = 3$ is odd so we use the LC criteria (B.10), i.e.,

$$a_3 > 0, ; a_1 > 0$$

and

$$\Delta_2 = det \begin{bmatrix} a_1 & a_3 \\ 1 & a_2 \end{bmatrix} = a_1 a_2 - a_3 > 0$$

These conditions for a cubic polynomial to be Hurwitz were first derived in the classical paper on *Governors* written by Maxwell in 1868 [33], and are sometimes referred to as *Maxwell's conditions*.

For the polynomial $p_2(s)$, $n = 4$ is even so we use the LC criterion (B.9), i.e.,

$$a_4 > 0, \ a_2 > 0$$

and

$$\Delta_1 = det[a_1] = a_1 > 0, \Delta_3 = det \begin{bmatrix} a_1 & a_3 & 0 \\ 1 & a_2 & a_4 \\ 0 & a_1 & a_3 \end{bmatrix} = a_1 a_2 a_3 - a_1^2 a_4 - a_3^2 > 0$$

Bibliography

[1] B.D.O. Anderson, "From Youla-Kucera to identification, adaptive and nonlinear control," *Automatica*, vol. 34, pp. 1485–1506, 1998.

[2] P.R. Bélanger, *Control Engineering:A Modern Approach*, Saunders College Publishing, Harcourt Brace and Co., Fort Worth, TX, 1995.

[3] H. S. Black, "Stabilized feedback amplifiers," U.S. Patent 2-102-671, 1927.

[4] V. Blondel, *Simultaneous Stabilization of Linear Systems*, Springer-Verlag, Berlin, 1994.

[5] S. P. Boyd and C. H. Barratt, *Linear Controller Design: Limits of Performance*, Prentice Hall, Englewood Cliffs, NJ, 1991.

[6] C. Carathéodory, *Theory of Functions of a Complex Variable, Vol. I*, Chelsea Publishing Company, New York, 1958.

[7] C. T. Chen, "Introduction to the linear algebraic method for control system design," *IEEE Control Systems Magazine*, vol. 7, pp. 36–42, 1987.

[8] C. T. Chen and C. S. Liu, "On control system design: A comparative study," *IEEE Control Systems Magazine*, vol.14, pp. 47–51, 1994.

[9] J.J. D'Azzo and C.H. Houpis, *Linear Control System Analysis and Design: Conventional and Modern*, McGraw-Hill, Inc., New York, 1995.

[10] P. Dorato, *Robust Control*, IEEE Press, New York, 1987.

[11] P. Dorato, C. Abdallah, and V. Cerone, *Linear-Quadratic Control: An Introduction*, Prentice Hall, Englewood Cliffs, NJ, 1995.

[12] P. Dorato, L. Fortuna, and G. Muscato, *Robust Control for Unstructured Perturbations–An Introduction*, vol. 168, Lecture Notes in Control and Information Sciences, Springer-Verlag, Berlin, 1992.

[13] P. Dorato, H. B. Park, and Yunzhi Li, "An algorithm for interpolation with units in H^∞, with applications to feedback stabilization," *Automatica*, vol. 25, pp.427–430, 1989.

[14] P. Dorato, "Quantified multivariate polynomial inequalities: The mathematics of (almost) all practical design problems," *6th IEEE Mediterranean Conference on Control and Systems*, Alghero, Italy, June 9–11, 1998.

[15] P. Dorato, D. Famularo, and C.T. Abdallah, "Analytic phase margin design," *IEEE Trans. Automat. Contr.*, scheduled for publication, February 2000 (Preliminary version published in Proceedings of the 6th IEEE Mediterranean Conference on Control and Systems, Alghero, Italy, June 9–11, 1998).

[16] R. C. Dorf and R. H. Bishop, *Modern Control Systems*, Addison-Wesley, Reading, MA, 1997.

[17] J. C. Doyle, B. A. Francis, and A. R. Tannenbaum, *Feedback Control Theory*, Macmillan, New York, 1992.

[18] G. F. Franklin, J. D. Powell, and A. Emami-Naeini, *Feedback Control of Dynamic Systems*, Addison-Wesley, Reading, MA, 1994.

[19] B. Friedland, *Control System Design: An Introduction to State-Space Methods*, McGraw-Hill, New York, 1986.

[20] F.R. Gantmacher, *Matrix Theory, Vol. II*, Chelsea Publishing Co., New York, 1959.

[21] E.A. Guillemin, *Synthesis of Passive Networks*, John Wiley, New York, 1957.

[22] S.M. Joshi, *Control of Large Flexible Space Structures*, vol. 131, Lecture notes in Control and Information Sciences, Springer-Verlag, Berlin, 1989.

[23] R. E. Kalman, "On the General Theory of Control Systems," *Proc. First International Congress, IFAC, Moscow, USSR*, 1960, pp. 481–492.

[24] L.H. Keel and S.P. Bhattacharyya, "Robust, fragile, or optimal?" *IEEE Trans. Automat. Control.*, vol. 42, pp. 1098–1105, 1997.

[25] H. Kimura, "Robust stabilization for a class of transfer functions," *IEEE Trans. Automat. Contr.*, vol. AC-29, pp. 788–793, 1984.

[26] B.C. Kuo, *Automatic Control Systems*, Prentice Hall, Englewood Cliffs, NJ, 1994.

[27] V. Kučera, "Stability of discrete linear feedback systems," *Proc. 6th IFAC World Congress*, Boston, MA, August 24–30, 1975.

[28] V. Kučera, *Discrete Linear Control*, John Wiley, New York, 1979.

[29] V. Kučera and F. J. Kraus, "FIFO stable control systems," *Proc. 12th IFAC World Congress, Vol. 1*, Sydney, Australia, 18–23 July, 1993, pp. 91–96.

[30] S. Malan, M. Milanese, and M. Taragna, "Robust analysis and design of control systems using interval arithmetic," *Automatica*, vol. 33, pp. 1364–1372, 1997.

[31] T.E. Marlin, *Process Control*, McGraw-Hill, New York, 1995.

[32] MATLAB *Control System Toolbox: Users Guide*, The MathWorks Inc., Natik, MA, 1996.

[33] J.C. Maxwell, "On Governors," *Proceeding of the Royal Society of London*, vol. 16, 1868, pp. 270–283 (Also reprinted in R. Bellman and R. Kalaba, *Selected Papers on Mathematical Trends in Control Theory*, Dover Publishers, New York, 1964).

[34] M. Morari and E. Zafiriou, *Robust Process Control*, Prentice Hall, Englewood Cliffs, NJ, 1989.

[35] G.C. Newton, Jr., L.A. Gould, and J.F. Kaiser, *Analytical Design of Linear Feedback Controls*, John Wiley, New York, 1957.

[36] N.S. Nise, *Control Systems*, Benjamin-Cummings, Menlo Park, CA, 1992.

[37] K. Ogata, *Modern Control Engineering*, Prentice Hall, Englewood Cliffs, NJ, 1997.

[38] C. L. Phillips and R. D. Harbor, *Feedback Control Systems*, Prentice Hall, Englewood Cliffs, NJ, 1996.

[39] J. R. Ragazzini and G. F. Franklin, *Sampled-data Control Systems*, McGraw-Hill, New York, 1958.

[40] R.S. Sánchez-Peña and Mario Sznaier, *Robust Systems: Theory and Applications*, John Wiley, New York, 1998.

[41] D.E. Seborg, T.F. Edgar, and D.A. Mellichamp, *Process Dynamics and Control*, John Wiley, New York, 1989.

[42] L.I. Smilen, "Interpolation on the real freqency axis," *IEEE Convention Record*, vol. 13, 1965, pp. 42–50.

[43] A. Tannenbaum, "Feedback stabilization of linear dynamical plants with uncertainy in the gain factor," *Int. J. Control*, vol. 32, pp. 1–16, 1980.

[44] J.G. Truxal, *Control System Synthesis*, McGraw-Hill, New York, 1955.

[45] M Vidyasagar, *Control System Synthesis: A Factorization Approach*, MIT Press, Cambridge, MA, 1985.

[46] M. Vidyasagar, "Statistical learning theory and randomized algorithms for control," *IEEE Control Systems*, vol. 18, pp. 69–85, 1998.

[47] B. Wie and D.S. Bernstein, "Benchmark problem for robust control design," *J. Guidance and Contr.*, vol. 15, pp. 1057–1059, 1992.

[48] D. C. Youla, J. J. Bongiorno Jr., and N. N. Lu, "Single-loop feedback-stabilization of linear multivariable dynamical plants," *Automatica*, vol. 10, pp. 159–173, 1974 (Also reprinted in [10]).

[49] D. C. Youla, H.A. Jabr, and J. J. Bongiorno Jr., "Modern Weiner-Hopf design of optimal controllers: Part II," *IEEE Trans. Automat. Contr.*, vol. AC-21, pp. 319–330, 1976 (Also reprinted in [10]).

[50] D. C. Youla and M. Saito, "Interpolation with positive real functions", *J. Franklin Inst.*, vol. 284, pp. 77–108, 1967.

[51] G. Zames and B. A. Francis, "Feedback, minimax sensitivity, and optimal robustness," *IEEE Trans. Automat. Contr.*, vol. AC-28, pp. 585–601, 1983 (Also reprinted in [10]).

Index